S0-BAS-253

Intermediate Algebra

Beta 1

by Lisa Healey

Intermediate Algebra
ISBN: 978-1-943536-13-9
Beta Edition 1.1 Winter 2017
© 2016, Chemeketa Community College. All rights reserved.

Chemeketa Press

Chemeketa Press is a nonprofit publishing endeavor at Chemeketa Community College. Working together with faculty, staff, and students, we develop and publish affordable and effective alternatives to commercial textbooks. All proceeds from the sale of this book will be used to develop new textbooks.

Publisher: Tim Rogers
Managing Editor: Steve Richardson
Production Editor: Brian Mosher
Book Editor: Steve Richardson, Matt "Math Slayer" Schmidgall
Design Manager: Kristen MacDonald
Cover Design: Kristi Etzel
Interior Design: Kristi Etzel
Layout: Kristi Etzel, Faith Martinmaas, Emily Evans, Cierra Maher, Candace Johnson

Additional contributions to the design and publication of this textbook come from the students and faculty in the Visual Communications program at Chemeketa.

Chemeketa Math Program

The development of this text and its accompanying MyOpenMath classroom includes the work of many faculty within the Chemeketa Math Program, including the following:

Development Team: Lisa Healey, Toby Wagner, Rick Rieman, Kyle Katsinis, Chris Nord
Reviewers: Ken Anderson, Tim Merzenich, Keith Schloeman
Early Adopter: Ken Anderson

Text Acknowledgment

This book has been developed using materials from *OpenStax College Algebra*, by OpenStax College, which have been made available under a Creative Commons Attribution 4.0 license and may be downloaded for free from cnx.org/contents/9b08c294-057f-4201-9f48-5d6ad992740d.

Printed in the United States of America.

Table of Contents

About Beta Editions

Chemeketa Press publishes textbooks using a software development model. Once a book is essentially ready for use, we send it to the classroom — even though it's still in the final stages of development. We do this so that students can start saving money as soon as possible and, more importantly, so that students and faculty will help us finish the book.

In exchange for a reduced price, we ask you to help us track down "bugs" in the book — typos, omissions, and other errors. We also ask you to let us know how well the book does its job and where it could do a better job. Whenever you have *anything* to share, please share it. You can talk with your professor, who will talk with us, or you can email the Press directly at **collegepress@chemeketa.edu.**

Thank you for participating in this exciting project that helps make college more affordable by reducing the price of textbooks.

CHAPTER 1

Graphs and Linear Functions

Toward the end of the twentieth century, the values of stocks of Internet and technology companies rose dramatically. As a result, the Standard and Poor's stock market average rose as well.

Figure 1 tracks the value of an initial investment of just under $100 over 40 years. It shows an investment that was worth less than $500 until about 1995 skyrocketed up to almost $1500 by the beginning of 2000. That five-year period became known as the "dot-com bubble" because so many Internet startups were formed. As bubbles tend to do, though, the dot-com bubble eventually burst. Many companies grew too fast and then suddenly went out of business. The result caused the sharp decline represented on the graph beginning around the year 2000.

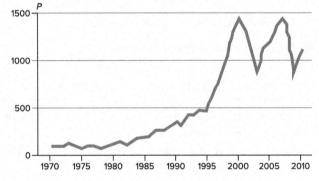

Figure 1.

Notice, as we consider this example, there is a relationship between the year and stock market average. For any year we choose we can estimate the corresponding value of the stock market average. Analyzing this graph allows us to observe the relationship between the stock market average and years in the past.

In this chapter, we will explore the nature of the relationship between two quantities.

1.1 Qualitative Graphs

OVERVIEW

In this section, we will see that, even without using numbers, a graph is a mathematical tool that can describe a wide variety of relationships. For example, there is a relationship between outdoor temperatures over the course of a year and the retail sales of ice cream. We can describe this relationship in a general way using a qualitative graph. As you study this section, you will learn to:

- Read and interpret qualitative graphs.
- Identify independent and dependent variables.
- Identify and interpret an intercept of a graph.
- Identify increasing and decreasing curves.
- Sketch qualitative graphs.

A. READING A QUALITATIVE GRAPH

Both qualitative and quantitative graphs can have two axes and show the relationship between two variables. We also read both from left to right — just like a sentence. The difference is that **quantitative graphs** have numerical increments (scaling and tick marks) on the axes, while **qualitative graphs** only illustrate the general relationship between the two variables.

▶ *Example 1*

Comparing Qualitative and Quantitative Graphs

Use the qualitative graph (Figure 1) and the quantitative graph (Figure 2) to answer the questions below.

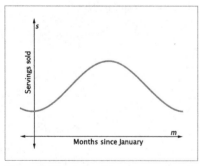

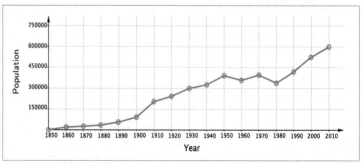

Figure 1. The sale of ice cream at Joe's Café (a qualitative graph).

Figure 2. The population of Portland, Oregon (a quantitative graph).

What does the qualitative graph tell us about ice cream sales at Joe's Café? Do we know how many servings were sold in June? What does the quantitative graph tell us about the population of Portland, Oregon? What was the population in 1930?

Solution

Ice cream sales are lowest at the beginning and at the end of the year and highest during the middle months. We cannot tell from this graph exactly how many servings are sold in any given month.

The population of Portland, Oregon has been increasing since 1850, except for a slight decrease in the 1950s and 1970s. The population in 1930 was about 300,000.

B. INDEPENDENT AND DEPENDENT VARIABLES

A qualitative graph is a visual description of the relationship between two variables. The graph tells a "story" about how one quantity is determined or influenced by another quantity. In the relationship between two variables — let's say p and t — if p depends on t, then we call p the **dependent variable** and t the **independent variable**.

▶ Example 2

Identifying Independent and Dependent Variables

Identify the independent variable and the dependent variable for each situation.

1. Let p represent the average price of a home in Salem, Oregon, and let t represent the number of years since 1990.

2. Let r represent the rate (in gallons per minute) that water is added to a bathtub and let m be the number of minutes it takes to fill the tub.

Solution

1. We say that the price p *depends on* or *is determined by* the year t. It is therefore the dependent variable. We would *not* say the year t depends on the average price of a home p. Time is independent of the price. Whether the average price goes up or down, time keeps on passing. So we call t the independent variable.

2. The rate of water flow determines how quickly the tub fills, so r is the independent variable. The number of minutes it takes to fill the tub depends on this rate, so m is the dependent variable.

Practice Set B

Determine the independent variable and the dependent variable for each situation. Turn the page to check your solutions.

1. Let m be the number of minutes since a cup of hot tea was poured, and let T be the temperature of the tea.

2. Let g be a student's exam score, and let s be the amount of time the student spent studying for the exam.

3. Let F be the outside temperature, and let c be the number of winter coats that a department store sells.

4. Let v be the resale value of a used car, and let a be the age of the car.

C. SKETCHING QUALITATIVE GRAPHS

When graphing, we always represent the independent variable along the horizontal axis, and we always represent the dependent variable along the vertical axis. In Figure 3, for example, we see that the height of a burning candle h is dependent on the number of minutes m since it has been lit. So the independent variable m is represented along the horizontal axis and the dependent variable h is represented along the vertical axis.

In Figure 3, you'll notice that the curve intersects both the horizontal and vertical axes. The point $(0, h)$ where the curve intersects the vertical axis is called the **vertical intercept** and the point $(m, 0)$ where the curve intersects the horizontal axis is called the **horizontal intercept**.

▶ *Example 3*

Interpreting Vertical and Horizontal Intercepts

Interpret the meaning of the intercepts of the graph in Figure 3.

Solution

The vertical-intercept or h-intercept on this graph represents the height of the candle in centimeters when it is first lit (when $m = 0$). The horizontal-intercept or m-intercept on this graph represents the time in minutes when the candle has been completely burned (when $h = 0$).

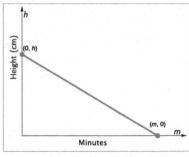

Figure 3. Height of a Burning Candle

If a curve goes upward from left to right (see Figure 4), we say the curve is **increasing**. If a curve goes downward from left to right (see Figure 5), we say the curve is **decreasing**. Some curves have both increasing and decreasing segments.

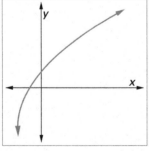

Figure 4. Increasing curve.

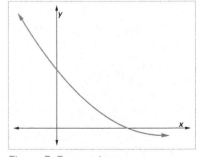

Figure 5. Decreasing curve.

▶ Example 4

Sketching a Qualitative Graph

A child climbs into a bathtub. After a few minutes of playing around in the water, the child gets out of the tub and pulls the plug so that all of the water drains away. Let W be the water level in the bathtub (in inches) at t minutes since the child climbed in. Sketch a qualitative graph that describes the relationship between the variables W and t.

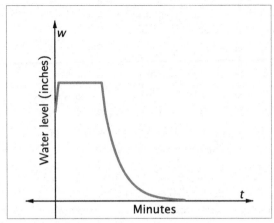

Figure 6. Water level over time in a bathtub

Solution

The water level W depends on t, the number of minutes that have passed, so in Figure 6 we let the vertical axis, representing the dependent variable, be the W-axis. We let the horizontal axis, representing the independent variable be the t-axis. The bathtub is full at the beginning of the situation, so we make sure that the W-intercept is well above the origin. When the child climbs into the tub, the water level rises a little so the curve will increase. When the child climbs out of the tub, the water level decreases a little. After the plug is pulled, the water level continues to decrease until the bathtub is drained. The t-intercept represents the time when the water level is zero.

Exercises

The population of a small town on the Oregon coast is described during the years between 2000 and the present. Let p be the population of the town at t years since 2000. For the following problems, match each of the figure 7 graphs to the scenario it describes.

1. The population increased steadily.

2. The population decreased steadily.

3. The population increased for 10 years then decreased.

4. The population remained constant.

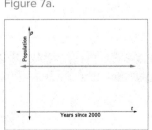

Figure 7a.

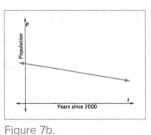

Figure 7b.

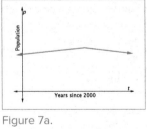

Figure 7c.

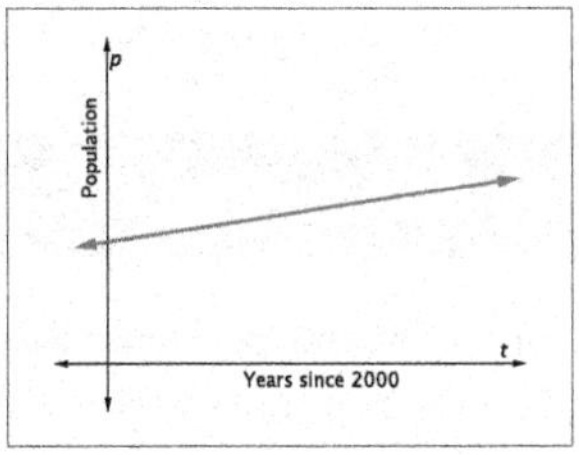

Figure 7d.

Alana goes for a 5-kilometer run each morning. Let d be the distance she has run t minutes after she begins. For the following problems, match each of the graphs in to the scenario it describes.

5. She runs at a steady pace the whole time.

6. She increases her speed the whole time.

7. She runs at a steady pace, then stops to rest, then continues at a slower pace.

8. She increases her speed for the first half of the run then decreases her speed.

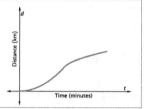

Figure 8a.

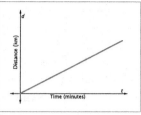

Figure 8b.

For the following exercises, identify the independent and dependent variables. Then sketch a qualitative graph that shows the relationship between the variables defined. Correct graphs may vary slightly and still accurately represent the given relationship.

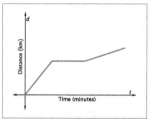

Figure 8c.

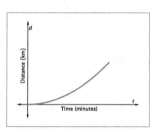

Figure 8d.

9. Let t be the amount of time (in minutes) it takes to read p pages in a novel.

10. Let c be the total cost (in dollars) of n lottery tickets.

11. Let T be the temperature (in degrees C) of a bowl of hot soup and let h the hours it is left uneaten on the dining room table.

12. Let s be the speed (in mph) of a train and let t be the amount of time (in hours) it takes to travel between two cities.

13. Let h be the height (in cm) of a sunflower, and let d be the days after it was planted as a seed.

14. Let h be the height of baseball (in feet) and let t be the time (in seconds) after a baseball bat hits it.

For the following exercises, sketch a qualitative graph that shows the relationship between the variables defined. Correct graphs may vary slightly.

15. Rodrigo left home, drove to another city, got gas, then continued driving to his cousin's house. Let g be the amount of gas (in gallons) in the gas tank at t minutes after he left home.

16. When Arianna was dieting, she lost weight quickly at first and then more slowly. She was then able to maintain a healthy weight. Let W be her weight (in pounds) for m months after she began the diet.

17. A plane flies from Portland to Los Angeles. Let a be the altitude (in feet) at t hours after takeoff.

18. Let h be the height (in feet) of a rubber ball at t seconds after a child throws it to the floor. It bounces several times and then stops.

Practice Set B — Answers

1. The independent variable is m, and the dependent variable is T

2. The independent variable is s, and the dependent variable is g

3. The independent variable is F, and the dependent variable is c

4. The independent variable is a, and the dependent variable is v

19. Pressure p (in pounds per square inch) is applied to a volume of gas in a closed container. As the pressure increases, the volume of gas v (in cubic cm) decreases.

20. Let h be the height (in meters) of a hot air balloon at t minutes after it launches. The balloon rises steadily at first, then stays a relatively constant height, then descends more slowly than it rose.

For the following exercises, let A be the amount of rain (in inches) that has fallen in t hours this afternoon. Sketch a qualitative graph for each of the following scenarios. Correct graphs may vary slightly.

21. The rain fell gently and then stopped. After a while, it began to rain hard.

22. The rain fell harder and harder.

23. The rain fell more and more gently.

24. The rain fell steadily all morning but the sun came out in the afternoon.

It finally stopped raining, and Mario went out for a walk. For the following exercises, let d be his distance from home (in meters) after leaving for t minutes. Sketch a qualitative graph for each of the following scenarios. The graphs are likely to vary slightly.

25. Mario walked quickly until he reached the park, then he turned and walked slowly home.

26. Mario walked slowly to the park, then turned and ran home.

27. Mario walked slowly at first, realized he forgot his hat, so then ran home. When he set out again, he kept a brisk pace to the park and back.

28. Mario walked steadily to the park, met a friend and stayed to talk for hours. His friend gave him a ride home in a car.

For the following exercises, write a scenario to match each of the following graphs. Make sure to define the variables x and y in your description.

29.

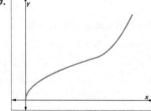

30.

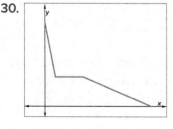

31.

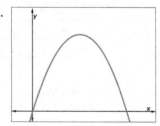

32.

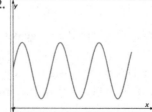

1.2 Functions

OVERVIEW

A jetliner changes altitude as the distance increases between it and the starting point of its flight. The weight of a growing child increases with time. In each case, one quantity depends on another. There is a relationship between the two quantities that we can describe, analyze, and use to make predictions. In this section, we continue studying such relationships using quantitative tools. We also introduce the concept of a function.

As you study this section, you will learn to:

♦ Know the meanings of relation, domain, range, and function.

♦ Determine whether an equation or table describes a function.

♦ Use the vertical line test to identify functions.

♦ Write domain and range as inequalities or in interval notation.

♦ Determine a function's domain and range from its graph.

♦ Use the Rule of Four to describe a function in multiple ways.

A. RELATIONS AND FUNCTIONS

A **relation** is a set of ordered pairs. The set consisting of the first components of each ordered pair is called the **domain** of a relation. The set consisting of the second components of each ordered pair is called the **range** of a relation.

Consider the following set of ordered pairs. The first numbers in each pair are the first five natural numbers. The second number in each pair is twice that of the first.

$$\{(1, 2), (2, 4), (3, 6), (4, 8), (5, 10)\}$$

The domain is $\{1, 2, 3, 4, 5\}$. The range is $\{2, 4, 6, 8, 10\}$.

A function f is a relation that assigns a single element in the range to each element in the domain. In other words, each input value has one and only one output value paired with it. In the example above, the relation that pairs the first five natural numbers with numbers double their values, is a function because each element in the domain, $\{1, 2, 3, 4, 5\}$, is paired with exactly one element in the range, $\{2, 4, 6, 8, 10\}$.

Domain and Range

The **domain** of a relation is the set of all values of the independent variable (**input** values). The **range** of a relation is the set of all values of the dependent variable (**output** values).

Note that each value in the domain is often labeled with the variable x. Each value in the range is often labeled with the variable y.

Function

A **function** is a relation where each value of the input variable leads to *exactly one* value of the output variable.

Figure 1 compares relations that are functions and not functions.

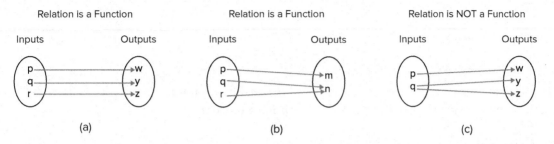

Figure 1. (a) This relationship is a function because each input is associated with a single output. In this case, each input is associated with a single output. (b) This relationship is also a function. Note that input q and r both give output n. (c) This relationship is not a function because input q is associated with two different outputs.

▶ Example 1

Determining Whether an Equation Describes a Function

Consider the relation that can be described by ordered pairs (x, y) such that $y = 2x - 1$. Is this relation a function?

Solution

This relation multiplies the input x by 2 and then subtracts 1 to obtain the output y. The domain of this relation is all real numbers, since we can do these operations to *any* real number. Let's try it with a few values from the domain.

$$
\begin{aligned}
y &= 2x - 1 \\
y &= 2(-3) - 1 = -6 - 1 = -7 \qquad \text{When } x = -3\,,\, y = -7 \\
y &= 2(0) - 1 = 0 - 1 = -1 \qquad \text{When } x = 0\,,\, y = -1 \\
y &= 2(4.5) - 1 = 9 - 1 = 8 \qquad \text{When } x = 4.5\,,\, y = 8
\end{aligned}
$$

Each input leads to only *one* output, so the relation $y = 2x - 1$ is a function.

▶ Example 2

Determining Whether an Equation Describes a Function

Consider the relation that can be described by ordered pairs (x, y) such that $y^2 = x$. Is this relation a function?

Solution

This relation takes the input x and matches it with an output y that will equal x when it is squared. The domain of this relation is the set of all real numbers greater than or equal to 0 ($x \geq 0$), since any real number squared is either positive or equal to 0.

Let's try it with a few values from the domain.

When $x = 1$, $y = 1$ or $y = -1$.

When $x = 4$, $y = 2$ or $y = -2$.

When $x = 9$, $y = 3$ or $y = -3$.

We can see that these input values each lead to *two* different output values. This violates the definition of a function, so the relation $y^2 = x$ is not a function.

▶ Example 3

Determining Whether a Table Describes a Function

In a particular math class, the overall percent score corresponds to a grade-point average. Figure 2 shows a possible rule for assigning grade points. Is grade-point average a function of the percent score?

Percent score	0–56	57–61	62–66	67–71	72–77	78–86	87–91	92–100
Grade-point average	0.0	1.0	1.5	2.0	2.5	3.0	3.5	4.0

Figure 2.

Solution

For any percent score earned, there is exactly one associated grade-point average, so the grade-point average is a function of the percent score. In other words, if we input any percent score, the output for that percent score is *one specific* grade point average.

▶ Example 4

Determining Whether a Relation Described Verbally is a Function

Determine whether the following descriptions are functions.

1. This relation assigns the student's age in years to each student ID number.

2. To each age in years, this relation assigns the ID number of the student(s) that are that age.

Solution

1. For this relation, the input is the student ID number and it determines the age of the student who has that ID number. So, this relation is a function because a student with a certain ID number can be *only one* age.

2. For this relation, the input is an age and it determines the ID number of a student who is that age. Since there will likely be *many* students with corresponding ID numbers who are the same age, this relation is not a function.

Practice Set A

Determine if the relations are functions. Explain how you know. Turn the page to check your solutions.

1. Ordered pairs (x, y) such that $y = \frac{x}{2}$

2. Ordered pairs (x, y) such that $y = \pm x$

3. Ordered pairs (t, p) such that $p = 15t$

4.

x	-3	-2	-1	0	1	2	3	4	5
y	21	20	14	9	3	8	15	20	26

5.

t	-3	0	1	3	-1	-3
h	16	24	38	31	29	13

6.

x	0	5	10	15	20	25
y	12	12	12	12	12	12

B. VERTICAL LINE TEST

We have seen in the examples above, that we can represent a relation using an equation, a table, or a verbal description. We can also represent a relation using a graph. Graphs display a great many input-output pairs efficiently. The visual information they provide often makes relationships easier to understand. By convention, graphs are typically constructed with the input values (the independent variable) along the horizontal axis and the output values (the dependent variable) along the vertical axis.

The Vertical Line Test

A relation is a function if and only if each vertical line intersects the graph of the relation at no more than one point. We call this requirement "passing the **vertical line test**".

An easy way to determine if a graph represents a function is to imagine a vertical line sweeping across the graph. A vertical line that hits a graph in *exactly one* point pairs that one input value of x with *exactly one* output value of y. In that case, the graph represents a function. On the other hand, when a vertical line hits a graph in *two or more* points, it pairs that input value of x with *two or more* output values of y. That graph does not represent a function. See Figure 3.

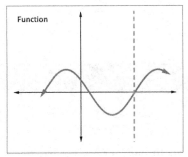

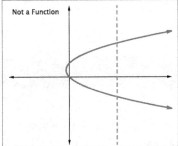

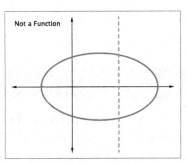

Figure 3.

▶ *Example 5*

Applying the Vertical Line Test

Which of the graphs in Figure 4 represent functions?

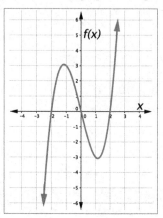
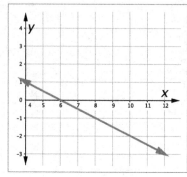
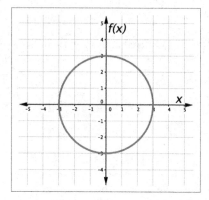

Figure 4.

Solution

For the graphs shown in parts (a) and (b), notice that any vertical line would pass through only one point. The graphs in (a) and (b) both pass the vertical line test and are therefore both functions. The third graph does not represent a function because — at most of the *x*-values — a vertical line would intersect the graph at more than one point, as shown in Figure 5.

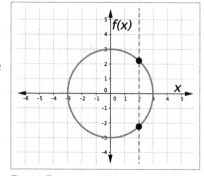

Figure 5.

Practice Set A — Answers

1. Function: Each input is paired with *only one* output.

2. Not a function: Each input is paired with *two* outputs. For example, if *x* = 2, then *y* = 2 and *y* = –2.

3. Function: Each input is paired with *only one* output.

4. Function: Each input is paired with *only one* output.

5. Not a function: The input *t* = –3 is paired with *two* outputs, namely *y* = 16 and *y* = 13.

6. Function: Each input is paired with *only one* output. It's okay that all the outputs are the same.

Practice Set B

Do the graphs below represent functions? Check your solutions on the following page.

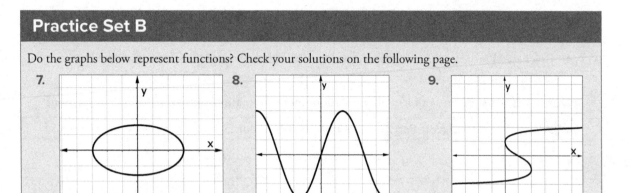

7.   8.   9.

C. DESCRIBING INTERVALS FOR DOMAIN AND RANGE

We can visualize the domain as a "holding area" that contains "raw materials" for a "function machine." The range is another "holding area" for the machine's products. Figure 6 illustrates this.

When the domain and range are all real numbers in an interval, then it is convenient to use **inequality notation** to define the domain and range. Inequalities use values, inequality symbols, and variables to describe a set of numbers. For example, if the domain is all real numbers less than 5, we could write the domain as the inequality $x < 5$. Recall the symbol < indicates less than a value and ≤ indicates less than or equal to a value. Similarly, the inequality symbol > indicates greater than a value and ≥ indicates greater than or equal to a value. As another example, if the range was all real numbers greater than -2 but no larger than 10, we could write the range as an inequality:

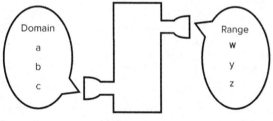

Figure 6.

$$-2 < y \leq 10$$

We can also describe domain and range in **interval notation**, which uses values within brackets to describe a set of numbers. In interval notation, we use square brackets [ ] when the set includes the endpoint and parentheses ( ) to indicate that the endpoint is either not included or the interval is unbounded. For example, if a person has $100 available to spend, the possibilities of how much she or he actually does spend would be the interval from 0 to 100, inclusive. In interval notation this is [0, 100]. If we knew the person was going to spend at least some money, then the interval would be (0, 100].

Here are the conventions of interval notation:

- ◆ The smallest number in the interval is written on the left, followed by a comma.

- ◆ The largest number in the interval is written on the right.

- Parentheses ( ) signify that an endpoint value is not included. On a graph, this is indicated by an open circle or an arrow.

- Brackets [ ] indicate that an endpoint value is included. On a graph, this is indicated by a closed circle (no arrow).

Figure 7 offers a summary of inequality notation, interval notation, and graphical representations:

Inequality	Interval Notation	Graph on a Number Line	Verbal Description
$a \leq x \leq b$	$[a, b]$	●————● $\quad$ a $\quad$ b	x is between a and b, including a and b
$a < x \leq b$	$(a, b]$	○————● $\quad$ a $\quad$ b	x is between a and b, including b but not a
$a \leq x < b$	$[a, b)$	●————○ $\quad$ a $\quad$ b	x is between a and b, including a but not b
$a < x < b$	(a, b)	 ○————○ $\quad$ a $\quad$ b	x is between a and b, not including a or b
$x \geq a$	$[a, \infty)$	 ●————→ $\quad$ a	x is greater than or equal to a
$x > a$	(a, ∞)	 ○————→ $\quad$ a	x is greater than a, not including a
$x \leq b$	$(-\infty, b]$	 ←————● $\quad$ b	x is less than or equal to b
$x < b$	$(-\infty, b)$	←————○ $\quad$ b	x is less than b, not including b
$x \neq a$	$(-\infty, a) \cup (a, \infty)$	 ←——○——→ $\quad$ a	all values except for a, or x cannot equal a
$-\infty < x < \infty$	$(-\infty, \infty)$	←————————→	all real numbers

Figure 7.

▶ *Example 6*

Describing Sets on a Number Line

Use inequalities and interval notation to describe the intervals in Figures 8 and 9.

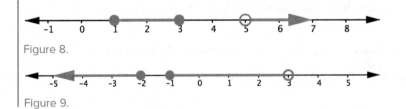

Figure 8.

Figure 9.

Solution

For Figure 8, the two intervals described are $1 \leq x \leq 3$ and $x > 5$. In interval notation, that is $[1, 3]$ and $(5, \infty)$. For Figure 9, the two intervals described are $x \leq -2$ and $-1 \leq x < 3$. In interval notation, that is $(-\infty, -2]$ and $[-1, 3)$.

▶ *Example 7*

Describing Domain and Range for Verbally Defined Functions

Use inequalities and interval notation to describe the domain and range of the functions. Assume the input variable is x and the output variable is y.

  a. A function where the inputs and outputs are all real numbers greater than but not equal to zero

  b. A function where the inputs are all real numbers, and the outputs are all real numbers between and including -5 and 5

  c. A function where the inputs are all real numbers greater than or equal to 0 but less than 33, and the outputs are all real numbers less than or equal to 7.

Solution

  a. Domain: $x > 0$ or $(0, \infty)$, Range: $y > 0$ or $(0, \infty)$

  b. Domain: $-\infty < x < \infty$ or $(-\infty, \infty)$, Range: $-5 \leq y \leq 5$ or $[-5, 5]$

  c. Domain: $0 \leq x < 33$ or $[0, 33)$, Range: $y \leq 7$ or $(-\infty, 7]$

D. USING A GRAPH TO FIND THE DOMAIN AND RANGE OF A FUNCTION

Another way to identify the domain and range of functions is by using graphs. Because the domain is the set of input values, the domain of a graph is represented on the horizontal axis. The range is the set of output values, which are represented on the vertical axis. Keep in mind that if the input-output pairs of the function continue beyond the portion of the graph we can see, the domain and range may be greater than the visible values.

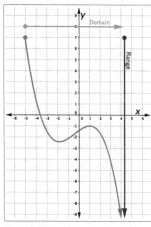

Figure 10.

As you can see in Figure 10, the graph extends horizontally from -5 to the right without bound, so the domain is $[-5, \infty)$. The vertical extent of the graph is all values 7 and below, so the range is $(-\infty, 7]$.

Remember when using interval notation, the smaller value is written on the left. For domains that will be the x-coordinate of the leftmost point on the graph. For ranges the smaller value will be the y-coordinate of the lowest point on the graph.

Practice Set B — Answers

7. No   **8.** Yes   **9.** No

▶ *Example 8*

Finding Domain and Range from a Graph

Find the domain and range of the function whose graph is shown in Figure 11.

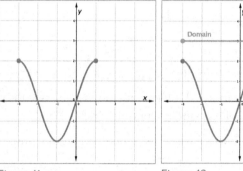

Solution

We can observe that the horizontal extent of the graph is –3 to 1. The closed circles indicate that we should include –3 and 1. So the domain of *f* is [–3, 1].

   The vertical extent of the graph is from –2 to 2, so the range is [–2, 2]. See Figure 12.

Figure 11.   Figure 12.

▶ *Example 9*

Finding Domain and Range from a Graph of Oil Production

Find the domain and range of the function whose graph is shown in Figure 13.

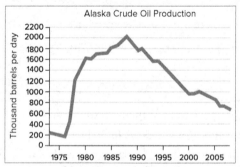

Figure 13. Alaskan crude oil production data from the U.S. Energy Information Administration.

Solution

The input quantity along the horizontal axis is "years," which we represent with the variable t for time. The output quantity is "thousands of barrels of oil per day," which we represent with the variable b for barrels. The graph may continue to the left and right beyond what is viewed, but based on the portion of the graph that is visible, we can determine the domain as $1973 \le t \le 2008$ and the range as approximately $180{,}000 \le b \le 2{,}010{,}000$.

   In interval notation, the domain is [1973, 2008], and the range is about [180000, 2010000]. For the domain and the range, we approximate the smallest and largest values since they do not fall exactly on the grid lines.

Practice Set D

Try the following. Turn the page to check your solutions.

10. Given Figure 14, identify the domain and range using interval notation.

11. Identity the domain and range of the relations graphed in Figure 15.

12. Which of the graphs in Figure 15 are functions?

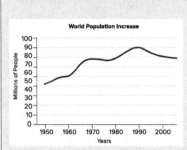

Figure 14.

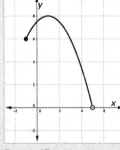 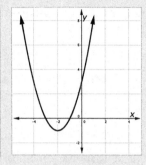 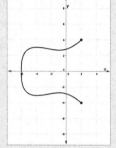

Figure 15.

E. RULE OF FOUR FOR FUNCTIONS

All of the functions that we will study in this course can be described in four ways: symbolically, graphically, numerically, and verbally. This is known as the **Rule of Four**.

Sometimes one of the four ways to describe a function may be more insightful or useful in a situation than another way. There may also be times when representing a function in multiple ways will be useful. You will benefit by learning to move easily between one way of describing a function and another.

Example 10 presents a simple function and how it is described by each of the four ways. Note that the table only shows some selected values for the function.

The Rule of Four for Functions

The description of the input-output pairs of a function can be

1. symbolic or algebraic (an equation)

2. verbal (words)

3. graphical (a graph)

4. numeric (a table).

Below is a simple function described by each of the four ways. The table only shows some selected values for the function.

1. $y = 2x + 1$	2. Multiply the input by 2, then add one to obtain the output.
3. 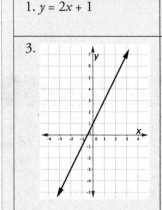	4. 

For table in cell 4:

x	y
-3	-5
-2	-3
-1	-1
0	1
1	3
2	5
3	7

▶ *Example 10*

Using the Rule of Four
to Describe a Function

A function squares the input and then subtracts 3 to obtain the output.

a.   Write an equation that matches the function description. Use x for the input variable and y for the output variable.

b.   Create a table of values to describe the function. Use $x = -3, -2, -1, 0, 1, 2, 3$ for the input values.

c.   Use the table to create a graph of the function.

d.   Use the graph to help you determine the domain and range of the function.

Solution

a. $y = x^2 - 3$

b.   The Figure 17 table shows some possible ordered pairs for the function.

c.   In Figure 18, we connect the points on our graph with a smooth curve since x can be any real number. We put arrows at each end of the curve for the same reason.

d.   Domain: $(-\infty, \infty)$, Range: $[-3, \infty)$

x	y
−3	6
−2	1
−1	−2
0	−3
1	−2
2	1
3	6

Figure 17.

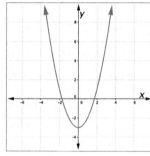

Figure 18.

Practice Set D — Answers

10. The domain is [1950, 2002] and the range is approximately [42,000,000, 90,000,000].

11. (a) Domain: [−1, 5), Range: (0, 8)
(b) Domain: (−∞, ∞), Range: (−2, ∞)
(c) Domain: [−6, 2], Range: [−4, 4]

12. Graphs (a) and (b) represent functions. Graph (c) does not pass the vertical line test, so it is not a function.

Exercises

1. What is the difference between a relation and a function?

2. What is the difference between the input and the output of a function?

3. Why does the vertical line test tell us whether the graph of a relation represents a function?

4. What is the difference between the domain and the range of a function?

For the following exercises, determine whether the relation represents y as a function of x.

5. $5x + 2y = 10$

6. $3y = x + 6$

7. $x = y^2 + 3$

8. $y = x^2$

9. $y = -2x^2 + 30x$

10. $2x + y^2 = 8$

11. $y = \frac{1}{x}$

12. $x = y^3$

For the following exercises, use the vertical line test to determine which graphs show relations that are functions.

13.
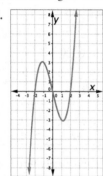

14.
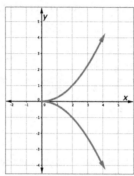

15.
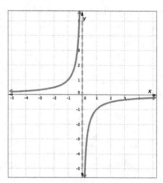

16.
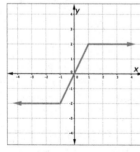

17.
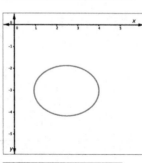

18.
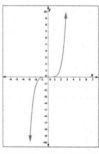

19.
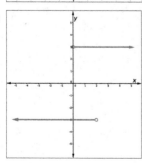

20.
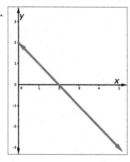

21.
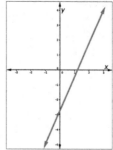

22. 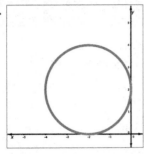   **23.**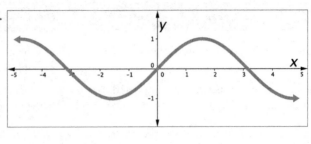

For the following exercises, determine whether the relation represents a function.

24. This relation begins with a person's social security number and pairs it with their date of birth.

25. This relation begins with a date of birth and pairs it with a social security number.

26. This relation assigns a height h to a rocket t seconds after the rocket is launched.

For the following exercises, determine if the relation represented in table form represents y as a function of x.

27.

x	5	10	15	20	25	30
y	3	8	14	21	29	38

29.

x	−4	−2	−4	0	2	4
y	−15	−20	−10	−5	0	5

28.

x	0	3	6	9	12	15
y	3	8	14	15	14	8

30.

x	−3	−2	−1	0	1	0
y	7	4	1	−2	−5	−8

For the following exercises, use the graph of the function to determine the function's domain and range. Write both as inequalities and in interval notation.

31.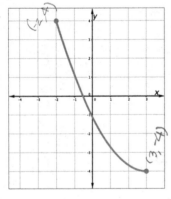

32.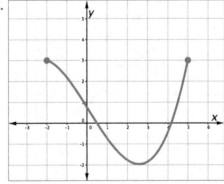

33.

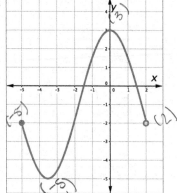

34.

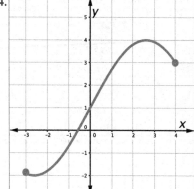

35.

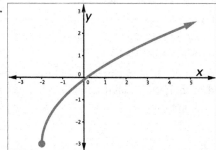

36.

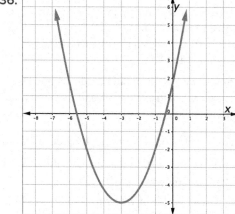

37.

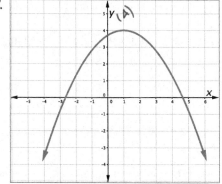

38.

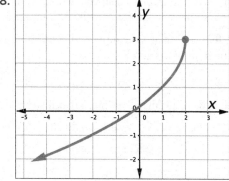

39.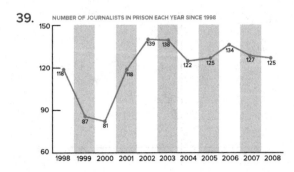

40.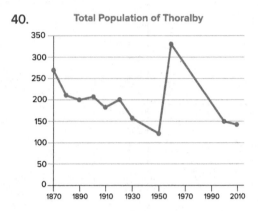

For the following verbally described functions, write the domain and range using interval notation.

41. A function where the inputs are all real numbers between and including 0 and 9.8, and the outputs are all real numbers between and including 0 and 225.

42. A function where the inputs are all real numbers greater than 0 and at most 30, and the outputs are all real numbers greater than 0 but less than 17.6.

43. A linear function where the inputs are all real numbers greater than –12, and the outputs are all real numbers less than –7.

44. A function where the inputs are all real numbers, and the outputs are all real numbers between and including –1 and 1.

For exercises 45 and 46, use your graphing calculator to help you do the following:

   a. Write an equation that matches the function description. Use *x* for the input variable and *y* for the output variable.

   b. Create a table of values to describe this function. Use *x* = –3, –2, –1, 0, 1, 2, 3 for the input values.

   c. Sketch a graph of the function.

   d. Use the graph to help you determine the domain and range of the function. Remember, the graph on your calculator screen will not show arrows.

45. A function squares the input and then adds 2 to obtain the output.

46. A function multiplies the input by 3 and then subtracts 4 to obtain the output.

For exercises 47 and 48, use your graphing calculator to help you do the following:

   a. Write a verbal description that matches the equation.

   b. Create a table of values to describe this function. Use *x* = –3, –2, –1, 0, 1, 2, 3 for the input values.

   c. Sketch a graph of the function.

   d. Use the graph to help you determine the domain and range of the function. Remember, the graph on your calculator screen will not show arrows.

47. $y = -2x + 5$

48. $y = x^3 - 1$

1.3 Finding Equations of Linear Functions

OVERVIEW

Imagine placing a plant in the ground one day and finding that it has doubled its height just a few days later. Although it may seem incredible, this can happen with some species of bamboo. These members of the grass family are the fastest-growing plants in the world. One species can grow nearly 1.5 inches every hour. In a twenty-four hour period, this bamboo plant grows about 36 inches — 3 feet! A constant rate of change, such as the change in height over time of this bamboo plant, is indicative of a linear function.

In Section 1.2, you learned that a function is a relation that assigns to every element in the domain exactly one element in the range. Linear functions are a specific type of function that can be used to model many real-world applications, such as plant growth over time. In this chapter, we will explore linear functions, their graphs, and how to relate them to data.

As you study linear functions, you will learn how to:

- ◆ Represent a linear function verbally, algebraically, graphically, and numerically.
- ◆ Determine whether a linear function is increasing, decreasing, or constant.
- ◆ Interpret slope as a rate of change.
- ◆ Build linear models from verbal descriptions.
- ◆ Build linear models from data in a table.
- ◆ Interpret the intercepts of a linear model.

A. REPRESENTING LINEAR FUNCTIONS

Many real world situations exhibit constant change over time. These situations can be represented with a **linear function**, which is a function with a constant rate of change. Consider, for example, the first commercial magnetic levitation (maglev) train in the world, the Shanghai MagLev Train. It carries passengers comfortably for a 30-kilometer trip from the airport to the subway station in only eight minutes.

Suppose a maglev train travels a long distance, maintaining a constant speed of 83 meters per second for a period of time once it is 250 meters from the station. The function describing the train's distance from the station at a given point in time is a linear function because the speed of the train is a constant rate of change. There are several ways to represent a linear function, including word form, slope-intercept equation form, tabular form, and graphical form.

In the examples below, we describe the train's motion as a function using each of these methods.

▶ *Example 1*

Representing a Linear Function in Verbal Form

For the train problem we considered above, write a sentence that may be used to describe the function relationship.

Solution

The train's distance from the station is a function of the time in seconds during which the train moves at a constant speed of 83 meters persecond. The train was 250 meters from the station when it began moving at this constant speed.

The speed is the rate of change. The rate of change for this example is constant, which means that it is the same between each input-output pair. As the time (input) increases by 1 second, the corresponding distance (output) increases by 83 meters. The train began moving at this constant speed at a distance of 250 meters from the station.

Another way to represent a linear function is to use an algebraic equation. One example is an equation written in slope-intercept form, $y = mx =+ b$, where x is the input value, m is the rate of change (slope), and b is the initial value of the dependent variable.

> ### Slope-Intercept Form of a Line
>
> Linear functions can be written in the **slope-intercept form** of a line
>
> $$y = mx + b$$
>
> where b is the initial or starting value of the function (when input, $x = 0$), and m is the constant rate of change, or slope of the function. The y-intercept is at $(0, b)$.

▶ *Example 2*

Representing a Linear Function in Slope-Intercept Form

Use the verbal description from Example 1 to write an equation in slope-intercept form that represents the motion of the train.

Solution

Let x, the input variable, represent the time in seconds that the train has been traveling at a constant speed, and let y represent the train's distance from the station. The rate of change m is 83 meters per second. The initial value of the dependent variable b is the original distance from the station, 250 meters. We can write this equation to represent the motion of the train.

$$y = 83x + 250$$

▶ *Example 3*

Representing a Linear Function in Tabular Form

For the motion of the train described in Examples 1 and 2, represent the relationship between the distance from the station and the time traveled in a table.

Solution

From the table in Figure 1, we can see that the distance changes by 83 meters for every 1 second increase in time.

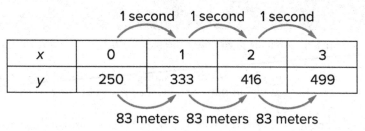

x	0	1	2	3
y	250	333	416	499

83 meters   83 meters   83 meters

Figure 1. Tabular representation of the linear function showing selected input and output values

▶ *Example 4*

Representing a Linear Function in Graphical Form

Graph the portion of the function, $y = 83x + 250$, that represents the motion of the train described in the previous examples.

Solution

Notice the graph in Figure 2 is a straight line. When we plot a linear function, the graph is always a straight line.

The rate of change, which is constant, determines the slant, or slope of the line. The point at which the input value is zero is the vertical intercept— also known as the *y*-intercept — of the line. We can see from the graph that the *y*-intercept in the train example is (0, 250) and represents the distance of the train from the station when it began moving at a constant speed.

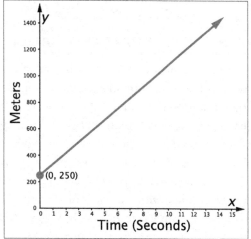

Figure 2. The graph of $y = 83x + 250$ for $x \geq 0$.

Note that the input represents time *after* a certain moment, so while positive real numbers are possible, negative real numbers are not possible for this example. The **practical domain** for this real-world linear function consists of non-negative real numbers represented by the interval $[0, \infty)$.

The domain and therefore the graph of the train in our example is restricted, but this is not always the case. Consider the graph of the line $y = 2x + 1$. Ask yourself what numbers can be put into the function; what is the domain of the function? The domain is comprised of all real numbers because any number may be doubled and then have one added to the product.

In general, the domain of all linear functions is $(-\infty, \infty)$. For all nonconstant linear functions, the rainge is also $(-\infty, \infty)$. However, when working with linear functions within the context of a real-world situation, we must stop and consider the limits on the input variable that give a practical domain.

Practice Set A

Use the description of employment at a trucking company to complete the exercises below. Turn the page to check your solutions.

Each week, the DiCicco Brothers Trucking Company pays its drivers $200 plus $18 per hour for every hour on the road. The Department of Transportation (DOT) limits the road time of employed drivers to 60 hours per week.

Let *x* represent the number of hours the employee spends driving in any week. Let *y* represent the employee's weekly pay in dollars.

1. Write a linear function in slope-intercept form to represent this situation.

2. Make a table of values for inputs $x = 0, 10, 20, 30, 40, 50$ hours.

3. Use a graphing calculator to graph the function. Make sure to use the values in your table to help you choose a good viewing window.

4. What is the practical domain of this function? What is the corresponding range?

B. SLOPE IS A RATE OF CHANGE

The linear function we used in the previous example increased over time, but not every linear function does. A linear function may be increasing, decreasing, or constant. For an increasing function, as with the train example, the output values increase as the input values increase. The graph of an increasing function has a positive slope. A line with a positive slope slants upward from left to right as in Figure 3a. For a decreasing function, the slope is negative. The output values decrease as the input values increase. A line with a negative slope slants downward from left to right as in Figure 3b. If the function is constant, the output values are the same for all input values so the slope is zero. A line with a slope of zero is horizontal as in Figure 3c.

Using Slope to Determine Increasing and Decreasing Functions

The slope determines if the function is an increasing linear function, a decreasing linear function, or a constant function.

$y = mx + b$ is an increasing function if $m > 0$.
$y = mx + b$ is a decreasing function if $m < 0$.
$y = mx + b$ is a constant function if $m = 0$.

Increasing linear function	Decreasing linear function	Constant function
(a)	(b)	(c)

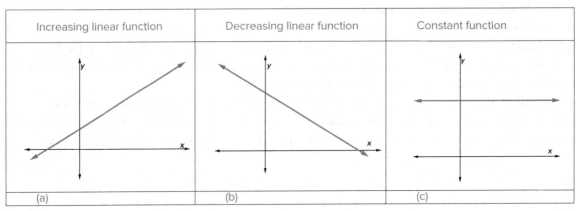

Figure 3.

▶ *Example 5*

Deciding Whether a Function Is Increasing, Decreasing, or Constant

Suppose that a teenager sends an average of 60 texts per day. For each of the following scenarios, find the linear function that describes the relationship between the input value and the output value. Then, determine whether the graph of the function is increasing, decreasing, or constant.

a.   The total number of texts a teen sends is considered a function of time in days. The input is the number of days, and output is the total number of texts sent.

b.   A teen has a limit of 500 texts per month in his or her data plan. The input is the number of days, and output is the total number of texts *remaining* for the month.

c.   A teen has an unlimited number of texts in his or her data plan for a cost of $50 per month. The input is the number of days, and output is the total cost of texting each month.

Solution

a. The function can be represented as $y = 60x$ where x is the number of days. The slope, 60, is positive, so the function is increasing. This makes sense because the number of texts increases each day.

b. The function can be represented as $y = -60x + 500$ where x is the number of days. In this case, the slope is negative so the function is decreasing. This makes sense because the number of texts *remaining* decreases each day and this function represents the number of texts remaining in the data plan after x days.

c. The cost function can be represented as $y = 50$ because the number of days does not affect the total cost. The slope is 0 so the function is constant.

In the examples we have seen so far, the slope was provided to us. However, we often need to calculate the slope given input and output values. Recall that given two values for the input, x_1 and x_2, and two corresponding values for the output, y_1 and y_2 — which can be represented by a set of points (x_1, y_1) and (x_2, y_2) — we can calculate the slope m.

The slope of a linear function can be calculated as follows:

$$m = \frac{\text{change in output (rise)}}{\text{change in input (run)}} = \frac{\Delta y}{\Delta x} = \frac{y_2 - y_1}{x_2 - x_1}$$

where x_1 and x_2 are input values and y_1 and y_2 are the corresponding output values. The units for slope are always described by $\frac{\text{units for the output}}{\text{units for the input}}$. Think of the units of a slope as "output units per input units." Examples of units for slope could be miles per hour, dollars per day, or pounds per cubic yard.

Given two input-output pairs for a linear function, we can calculate m, the slope or rate of change, and interpret it with the following steps:

1.   Determine the units for input and output values and write the data given as two points.

2.   Calculate the slope according to the formula $m = \frac{y_2 - y_1}{x_2 - x_1}$.

3.   Interpret the slope as the change in output values per unit of the input value.

▶ *Example 6*

Using Population Change to Find the Slope of Linear Function

The population of McMinnville, Oregon, increased from 26,695 to 33,131 between 2000 and 2013. Find the slope of the linear function that would represent the population growth if we assume the rate of change was constant from 2000 to 2013.

Solution

The input values would be the given years because time is the independent variable. The output values would be the number of people because the population depends on the year. The data can be represented as ordered pairs: (2000, 26695) and (2013, 33131).

We calculate the slope in this way:

$$m = \frac{y_2 - y_1}{x_2 - x_1}$$

$$= \frac{33131 - 26695}{2013 - 2000}$$

$$= \frac{6436}{13}$$

The numerator represents an increase of 6,436 people. The denominator represents a span of 13 years.

$$\approx 495 \frac{\text{people}}{\text{year}}$$

In this situation, it makes sense to round to the nearest whole person (per year).

So, we interpret the slope to mean that between 2000 and 2013, the population of McMinnville increased by approximately 495 people per year.

Practice Set A — Answers

1. $y = 18x + 200$

2.

x (hours)	y (dollars)
0	200
10	380
20	560
30	740
40	920
50	1100

3.

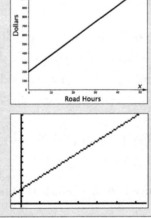

4. The practical domain represents hours on the road so only zero and positive values make sense. The DOT limits the number of road hours per week to 60. So the domain is [0, 60]. The corresponding range is the dollars earned from a minimum of $200 (when $x = 0$) to a maximum of $1280 (when $x = 60$). So the range is [200, 1280].

Practice Set B

Each situation below can be described by a linear function. Find and interpret the slope of the function as a rate of change. Turn the page to check your solutions.

5. The population of Gervais, Oregon increased from 977 people in 1990 to 2,477 people in 2010. Assume the rate of change was constant during those years.

6. A backyard swimming pool is being drained at a constant rate. After 3 hours there are 8450 gallons of water remaining in the pool and after 10 hours there are 5650 gallons.

C. FINDING AN EQUATION OF A LINEAR MODEL USING DATA FROM WORDS

We can use linear functions to model real-world situations involving quantities with a constant rate of change — whether or not that rate of change is always perfectly constant. Functions that describe such situations are called **linear models**. As long as we know or can figure out the initial value and the rate of change of a linear function, we can solve many different kinds of real-world problems.

When building linear models to solve problems, we use the slope-intercept form: $y = mx + b$. The slope m represents the rate of change in the problem, while the y-intercept b represents a starting value.

▶ *Example 8*

Using a Linear Function to Model the Number of Songs in a Music Collection

Marcus currently has 200 songs in his music collection. Every month, he adds 15 new songs. Write a linear model for the number of songs, y, in his collection as a function of time, x, the number of months. How many songs will Marcus own at the end of one year?

Solution

The initial value for this function is 200 because he currently owns 200 songs, so when $x = 0$, $y = 200$, which means that the y-intercept is $(0,200)$ and $b = 200$.

Since Marcus adds 15 new songs to his collection each month, the slope of the line $m = 15$, representing an increase of 15 songs per month. We can substitute the initial value and the rate of change into the slope-intercept form of a line.

$$\begin{aligned} y &= mx + b \\ y &= 15x + 200 \end{aligned}$$

We can write the formula $y = 15x + 200$ to model the number of songs in Marcus's music collection after x months.

With this formula, we can then predict how many songs Marcus will have at the end of one year (12 months). In other words, we can evaluate the function at $x = 12$.

$$\begin{aligned} y &= 15(12) + 200 \\ &= 180 + 200 \\ &= 380 \end{aligned}$$

Marcus will have 380 songs in 12 months.

The previous example was simplified by the fact that the data given could directly be interpreted as the slope and the y-intercept of the linear model. However, sometimes the data given is in the form of two input-output pairs. In that case we need to use the steps outlined below to find a linear model.

Given two input-output pairs for a linear model, we can find the slope-intercept form of the line with the following process:

1. Determine the units for input and output values and write the data given as two points.

2. Calculate the slope according to the formula $m = \frac{y_2 - y_1}{x_2 - x_1}$.

3. Using the slope-intercept form of a line $y = mx + b$, substitute your answer from step 2 in for m and substitute the coordinates of one of your points from step 1 in for values of x and y.

4. Solve the equation for b.

5. Beginning again with slope-intercept form, substitute the values for m and b into the formula, leaving x and y as variables.

▶ Example 9

Using a Linear Function to Model Salary Based on Commission

As an insurance salesperson, Rosa earns a base salary plus commissions on new policies. Therefore, Rosa's weekly income, y, depends on the number of new policies, x, she sells during the week. Last week she sold 3 new policies and earned $760 for the week. The week before, she sold 5 new policies and earned $920. Find a linear function to model this data. Then interpret the meaning of the slope and y-intercept of the equation. Finally, determine Rosa's weekly income if she sells 10 new policies in a week.

Solution

The given information gives us two input-output pairs, (3, 760) and (5, 920). We start by finding the slope of the linear model.

Practice Set B — Answers

5. The data can be written as ordered pairs: (1990, 977) and (2010, 2477). The slope is

$$m = \frac{2477 - 977}{2010 - 1990}$$
$$= \frac{1500}{20}$$
$$= 75 \text{ people per year}$$

The population of Gervais increased by 75 people per year between 1990 and 2010.

6. The data can be written as ordered pairs: (3, 8450) and (10, 5650). The slope is

$$m = \frac{5650 - 8450}{10 - 3}$$
$$= \frac{-2800}{7}$$
$$= -400 \text{ gallons per hour}$$

The slope is negative and indicates that the water is *draining* out of the pool at a rate of 400 gallons per hour.

$$m = \frac{920 - 760}{5 - 3}$$

$$= \frac{160}{2}$$

$$= 80$$

Keeping track of units can help us interpret this quantity. Income increased by $160 when the number of policies increased by 2, so the rate of change is $80 per policy. Therefore, the slope tells us that Rosa earns a commission of $80 for each policy sold each week.

We can now use the form $y = mx+b$ to solve for b.

$y = 80x + b$	Substitute the value of the slope $m = 80$.
$760 = 80(3) + b$	Using one ordered pair (3, 760), we substitute $x = 3$, and $y = 760$.
$760 = 240 + b$	Simplify.
$520 = b$	Subtract 240 from both sides of the equation to solve for b.

The value of b is the starting value for the function and represents Rosa's income when $x = 0$, or when no new policies are sold. We can interpret this as Rosa's base salary for the week, which does not depend upon the number of policies sold. We can now write the final equation:

$$y = 80x + 520$$

Our final interpretation is that Rosa's base salary is $520 per week and she earns an additional $80 commission for each policy sold. If Rosa sells 10 new policies in a week, the input variable $x = 10$. Substituting this value in our formula we get:

$$y = 80(10) + 520$$

$$= 800 + 520$$

$$= 1320$$

So if Rosa sells 10 new policies in a week, her salary will be $1320.

Practice Set C

7. Bernardo starts a company in which he incurs a fixed cost of $1,250 per month for the overhead, which includes expenses like his office rent. His production costs are $37.50 per item. Write a linear function in slope-intercept form to model this situation where y is the total cost for x items produced in a given month. What is the company's cost if Bernardo produces 100 items in a month?

8. Walter's Well and Water Company is drilling a well for a customer. On Tuesday, after 5 hours, the drilling equipment has reached a depth of 50 feet, after 8 hours, they have reached a depth of 72.8 feet. Write a linear function in slope-intercept form that models this situation where y is the depth of the well after x hours. Interpret the meaning of the slope and y-intercept.

Turn the page to check your solutions.

D. FINDING AN EQUATION OF A LINEAR MODEL USING DATA FROM A TABLE

At the beginning of this section we saw that a linear function can also be represented by a table of values. In a real-world situation we can use the table of values to find the slope and y-intercept of a linear model.

▶ *Example 10*

Using a Table of Values to Write an Equation for a Linear Model

Figure 4 relates y, the number of dollars in a savings account to time x in weeks. Use the table to write a linear equation.

x (weeks)	0	2	4	6
y (dollars)	1000	1080	1160	1240

Figure 4.

Solution

We see from the table that the initial value for savings account is 1000, so $b = 1000$. Rather than solving for m, we can tell from looking at the table that the value increases by 80 every 2 weeks. This means the rate of change is 80 dollars per 2 weeks, which can be simplified to 40 dollars per week.

$$y = 40x + 1000$$

If we did not notice the rate of change from Figure 4 we could still solve for the slope using any two points from the table. For example, using (2, 1080) and (6, 1240)

$$m = \frac{1240 - 1080}{6 - 2}$$
$$= \frac{160}{4}$$
$$= 40$$

The initial value was provided in the table in the previous example, but sometimes it is not. If you see an input of 0, then the initial value would be the corresponding output. If the initial value is not provided because there is no value of input on the table equal to 0, find the slope, substitute one coordinate pair and the slope into $y = mx + b$, and solve for b.

Practice Set C — Answers

7. The total cost for x items produced in a month is given by $y = 37.50x + 1250$. If Bernardo produces 100 items in a month, his monthly cost is found by substituting 100 for x.

$$y = 37.5(100) + 1250$$
$$= 5000$$

So his monthly cost would be $5,000.

8. The data given can be written as input-output pairs (5, 50) and (8, 72.8). The slope between these two points $m = 7.6$ represents the rate of drilling, specifically 7.6 feet per hour. Solving for the y-intercept gives $b = 12$, which represents the depth of the well when the company began drilling on Tuesday. The linear function $y = 7.6x + 12$ models this situation.

Practice Set D

9. A new plant food was introduced to a young tree to test its effect on the height of the tree. Figure 5 shows the height of the tree, in centimeters, for some specific number of months since the measurements began. Write a linear function to model this situation. Let y equal the height of the tree x number of months since the start of the experiment.

x (months)	0	1	2	3	4
y (cm)	17	25.5	34	42.5	51

Figure 5.

10. Use your equation from the previous problem to estimate the height of the tree after 12 months, assuming it continues to grow at the same rate.

When you're finished, turn the page to check your solutions.

E. USING AND INTERPRETING THE INTERCEPTS OF A LINEAR MODEL

Some real-world problems provide the y-intercept, which is the initial value, or the y value when $x = 0$. Once an equation for a linear model is known, the x-intercept can be calculated. The x-intercept of the function is the value of x when $y = 0$. It can be found by solving the equation $0 = mx + b$.

▶ *Example 11*

Hannah plans to pay off a generous, no-interest loan from her parents. Her loan balance is \$4,000. She plans to pay \$250 per month until her balance is \$0.

   a. Write an equation to model this situation.

   b. Find and interpret the x-intercept of the model.

   c. Find the practical domain of the function.

Solution

a. The b-value is the initial amount of her debt, or \$4,000. The rate of change, or slope, is -250 dollars per month. We can use the slope-intercept form and the given information to develop a linear model.

$$y = mx + b$$
$$y = -250x + 4000$$

b. Replace y with 0, and solve for x to find the x-intercept.

$$0 = -250x + 4000$$
$$-4000 = -250x \qquad \text{Subtract 4000 from both sides.}$$

$$16 \;=\; x \qquad\qquad \text{Divide both sides by } -250.$$

$$x \;=\; 16 \qquad\qquad \text{Rewrite the equation with the variable on the left.}$$

The x-intercept is (16,0) and represents the number of months it takes Hannah to reach a balance of $0. It will take her 16 months to pay off her loan.

c. The domain of this function refers to the number of months that Hannah pays off her loan. In this case, it doesn't make sense to talk about input values less than zero. We have determined in part b that it will take her 16 months to pay off the loan. So the practical domain is [0, 16].

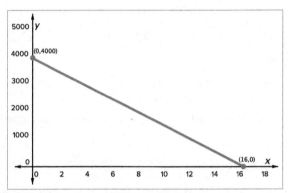

Figure 6. The graph of the function $y = -250x + 4000$.

Figure 6 presents a graph of the linear model for Example 11. The model is a decreasing linear function because the slope is negative and the curve decreases as we read the graph from left to right. Notice the location of the x- and y-intercepts on the graph.

Exercises

1. Terry is skiing down a steep hill. Terry's elevation, y, in feet after x seconds is given by $y = -70x + 3000$.

   a. Write a complete sentence describing Terry's starting elevation and how it is changing over time.

   b. Make a table of values to represent the function. Let $x = 0, 10, 20, 30, 40$.

   c. Sketch a graph of the function. You can use your graphing calculator to check your work.

2. The amount of water remaining in a water tank, y, in gallons after x minutes is given by $y = -3.5x + 1400$.

   a. Write a complete sentence describing the initial amount of water in the tank and how it is changing over time.

   b. Make a table of values to represent the function. Let $x = 0, 10, 20, 30, 40$.

   c. Sketch a graph of the function. You can use your graphing calculator to check your work.

3. A boat that was anchored 25 miles from a marina begins sailing directly away from it at 8 miles per hour.

   a. Write an equation for the distance of the boat y from the marina after x hours.

   b. Make a table of values to represent the function. Let $x = 0, 1, 2, 3, 4$.

   c. Sketch a graph of the function. You can use your graphing calculator to check your work.

4. Amanda has $370 in her savings account and she is adding $120 per month.

   a. Write an equation for the balance in the account y after x months.

   b. Make a table of values to represent the function. Let $x = 0, 3, 6, 9, 12$.

   c. Sketch a graph of the function. You can use your graphing calculator to check your work.

For the following exercises, determine whether each function is increasing or decreasing.

5. $y = 14x + 30$

6. $y = 5x + 16$

7. $y = 7 - 2x$

8. $y = 28 - 3x$

9. $y = -17.3x + 41.5$

10. $y = -22.48x + 1.92$

11. $y = \frac{1}{2}x - 3$

12. $y = \frac{1}{4}x - 5$

13. $y = -\frac{2}{5}x - \frac{1}{8}$

14. $y = -\frac{3}{7}x + \frac{1}{7}$

For the following exercises, assume the situation represents a linear function. Find and interpret the slope of the line that models the data.

15. Two months after its grand opening, a new museum has had a total of 489 visitors. After 6 months the total number of visitors to the museum is 1490.

16. A group of hikers sets out a long walk. After 0.25 of an hour, they have gone 1.2 miles. After 0.75 of an hour, they have traveled 3.6 miles.

17. A scuba diver descends to the bottom of a lake. After 30 seconds the diver is 137 feet from the bottom of the lake and after 2 minutes the diver is just 2 feet from the bottom of the lake.

18. Money is dispensed from a retirement account such that after 5 years there is $165,000 left in the account and after 10 years there remains $80,000.

For the following exercises, write a linear function in slope-intercept form that models the situation. Interpret the meaning of the slope and y-intercept in the context of the problem. Then, find the value of the function that satisfies the follow-up question.

19. A plumber charges a flat rate of $120 to make a house call and charges $42.50 per hour for labor. Let y be the total cost of the house call plus labor and let x be the number of hours worked. How much does it cost if the plumber works for 5.5 hours?

20. A hot air balloon is hovering 100 meters above ground and begins to ascend at a rate of 14 meters per second. Let y be the height of the balloon in meters x seconds after it begins to ascend. How high is the hot air balloon after 30 seconds?

21. Grain is pumped out of a full grain silo into a barge at a rate of 13 cubic meters per minute. The silo holds 4500 cubic meters of grain. Let y = cubic meters remaining in the silo after x minutes. How many cubic meters of grain remain in the silo after 4 hours?

22. Stock value per year in a certain company has been steadily decreasing in value by $1.80 per year for the

last 10 years. Ten years ago the value of the stock was $39.55. Let y be the value of the stock x years after it began to decrease. What was the value of the stock 7 years after it began to decline?

23. A cable lowers some equipment into a mine shaft. After 5 minutes the equipment is 82 meters from the bottom of the shaft and after 15 minutes the equipment is 46 meters from the bottom of the shaft. Let y = the distance to the bottom of the shaft in meters after x minutes. How far from the bottom of the shaft is the equipment after 20 minutes?

24. Sonny has entered a reading contest at the local library. After reading two books, he has read 330,000 words of text. After 5 books, he has read 825,000 words. Write a linear function to model his progress where x is the number of books read and y is the total number of words read. How many books will he need to read to reach a total of 10,000,000 words?

Practice Set D — Answers

9. $y = 8.5x + 17$

10. $y = 8.5(12) + 17 = 119$, so After 12 months the tree will be 119 cm tall.

25. Marco is farming gold coins to buy a new set of armor in an online game. After 3 hours, he has a total of 315 gold coins. After 8, he has 540. Create a linear function in slope-intercept form that models how many gold coins he has, where x is the number of hours he has spent farming and y is his total amount of gold coins. How long will he have to farm in order to have the 855 gold coins he needs?

26. Bianca is analyzing 220 megabits of data from the Mars Rover, *Curiosity*. After 90 minutes, she has 160 megabits left to analyze. Create a linear function where y is the number of megabits remaining to be analyzed and x is the number of minutes that have passed. How long will it take her to analyze all 220 megabits of data?

For the following exercises, find a linear equation that models the data.

27.

x	0	1	2	3	4
y	5	−7	−19	−31	−43

31.

x	−2	2	6	10	14
y	36	30	24	18	12

28.

x	0	2	4	6	8
y	−3	7	17	27	37

32.

x	4	6	8	10	12
y	13	18	23	28	33

29.

x	0	5	10	15	20
y	−5	10	25	40	55

33.

x	1	4	7	10	13
y	7.2	10.8	14.4	18	21.6

30.

x	0	3	6	9	12
y	184	154	124	94	64

34.

x	−5	5	15	25	35
y	14	−22	−58	−94	−130

For the following exercises, find and interpret the x and y-intercepts of each equation.

35. Freddie plays poker with his friends every Friday night. The following equation models the amount of money he has remaining after x rounds of Texas Hold 'Em. $y = -15x + 120$.

36. Sun-Mi has saved some spending money for food and activities while on vacation. The following equation models the amount of money she has remaining after x days. $y = -53x + 490$

For each of the following exercises, answer all four questions.

37. The number of gallons of water remaining in a watering trough x minutes after it begins draining is given by $y = -12x + 350$.

    a. Interpret the slope and y-intercept of this equation in the context of the problem.

    b. How many gallons remain in the trough after 14 hours?

    c. When will there be only 100 gallons remaining in the trough?

    d. How long does it take for the trough to be empty (0 gallons)?

38. A hiker descends from a mountain top. Her altitude above sea level in feet x hours after she begins her descent is given by $y = -260x + 1400$.

    a. Interpret the slope and y-intercept of this equation in the context of the problem.

    b. What is the hiker's altitude after 2.5 hours?

    c. When will the hiker be at an altitude of 555 feet?

    d. How long does it take the hiker to descend to sea level (0 feet)?

1.4 Using Linear Functions to Model Data

OVERVIEW

A professor is attempting to identify trends among final exam scores. His class has a mixture of students, so he wonders if there is any meaningful relationship between their ages and the final exam scores. One way for him to analyze the scores is by creating a diagram that relates the age of each student to the exam score received. In this section, we will examine one such diagram known as a scatter plot. Models such as scatter plots can be extremely useful for analyzing relationships and making predictions based on those relationships.

As you study this section, you will learn how to:

- Draw and interpret scatter plots.

- Distinguish between linear and nonlinear relations.

- Find a line of best fit by hand.

- Use a graphing utility to find a linear regression.

- Fit a regression line to a set of data and use the linear model to make predictions.

- Interpret the intercepts of a model.

- Determine when model breakdown occurs.

A. SCATTER PLOTS AND LINEAR MODELS

A **scatter plot** or scattergram is a graph of plotted points. A scatter plot can give us useful information about the relationship between two sets of data.

If the scattergram shows a relationship that is linear or nearly linear, we can write a linear equation to model the relationship. We can then use our equation to help us make predictions and draw conclusions.

Of course, not all relationships can be represented by linear models. The professor mentioned in this section's overview, who wants to see if a student's age is meaningfully related to their final exam score, can create a scatter plot of the data. See Figure 1.

You'll notice this scatter plot does not indicate a linear relationship. In fact the points do not appear to follow any trend. In other words, there does not appear to be a meaningful relationship between the age of the student and the score on the final exam.

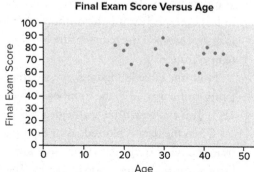

Figure 1. A scatter plot relating age and final exam scores

To determine whether a set of data is linearly related or nearly linearly related, we look at the scatter plot of the ordered pairs. If a set of data has a positive linear relationship, it can be modeled with a line having a positive slope. From Figure 2, we see that if a set of data has a negative linear relationship, it can be modeled with a line having a negative slope. We can also see that not all data can be modeled with a linear function.

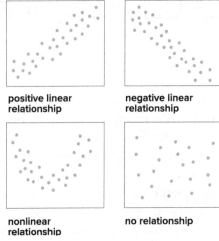

positive linear relationship   negative linear relationship

nonlinear relationship   no relationship

Figure 2. Scatter plots may or may not show a nearly linear relationship.

▶ *Example 1*

Using a Scatter Plot to Investigate Cricket Chirps

Figure 3 shows the number of chirps made by a cricket during a 15-second interval, for several different air temperatures, in degrees Fahrenheit. Make a scatter plot of this data, and determine whether the data appears to be linearly related.

Temperature (°F)	52	53	57	61	66	68	70.5	72	73.5	80.5
Chirps (in 15-second interval)	18.5	23	21.5	27	33	31	35	35	37	44

Figure 3. Air Temperatures vs. Cricket Chirps

Solution

Recall from Section 1.2 that data in a table represent input-output pairs, which in turn can be written as ordered pairs. Temperature is the independent variable in this example because the number of cricket chirps may *depend upon* the temperature but will not *determine* temperature.

From the table in Figure 3, we see that when the temperature was 52°F, the number of cricket chirps was 18.5. This can be written as the ordered pair (52, 18.5).

When the data is plotted, as in Figure 4, we can see from the trend in the data that the number of chirps increases as the temperature increases. The trend appears to be roughly linear, with a positive slope , though certainly not perfectly so.

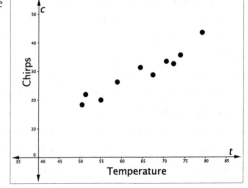

Figure 4. Scatter Plot of Temperature (°F) vs. Cricket Chirps (in 15 second interval).

Creating Scatter Plots on a Graphing Calculator

We can also use our graphing calculators to make a scatter plot.

Step 1: Press STAT and ENTER to access Lists 1 and 2, L_1 and L_2. To clear previous data, arrow up to the list title and press CLEAR and ENTER.

Step 2: Enter the input values into L_1 and the output values into L_2, as in Figure 5.

Step 3: Access the Stat Plot above the Y= button. ENTER to select Plot 1 and ENTER again to turn the plot on. Make sure the other icons on this screen match the selections in Figure 6.

Step 4: To view the scatter plot, Figure 7, select ZOOM 9 for a viewing window that matches the data.

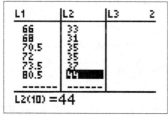

Figure 5.

Figure 6.

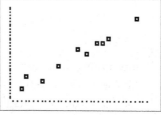

Figure 7. Example 1 scatter plot.

B. APPROXIMATING LINES OF BEST FIT

Once we recognize that a set of data is nearly linear, we can try to find the equation of a linear function that "best fits" the description of the data. One way to approximate our linear function is to sketch the line that seems to come closest to most data points while following the trend of the data. We call this a trend line or a **line of best fit**. See Figure 8 for an example.

To find the equation of our line, we can follow the steps outlined in Section 1.3 for writing an equation in slope-intercept form using two points. However, not all data points are created equal! Make sure to choose two points that are on or near your trend line.

▶ *Example 2*

Finding the Equation of an Approximate Line of Best Fit

Use the data in the table in Figure 3 from Example 1 and the trend line drawn in Figure 8 to find the equation in slope-intercept form of a linear model that fits the data. Interpret the slope of the line in the context of the problem.

Solution

Two data points appear to be on or near the line drawn in Figure 5. Checking our data table, we find the coordinates of these points are (61, 27) and (73.5, 37).

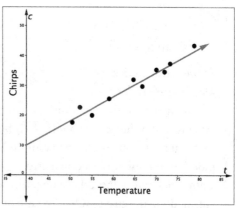

Figure 8. A line of best fit follows the trend of the data.

Now we calculate the slope of the line:

$$m = \frac{37 - 27}{73.5 - 61}$$

$$= \frac{10}{12.5}$$

$$= 0.8$$

The output units are chirps, while the input units are °F. So we can interpret the slope to mean there is an increase of 0.8 (almost 1) chirp per increase of 1°F. We can now solve for the y-intercept $(0, b)$.

$y = 0.8x + b$	Substitute the value of the slope $m = 0.8$
$27 = 0.8(61) + b$	Using one of our ordered pairs $(61, 27)$, we substitute $x = 61$, and $y = 27$.
$27 = 48.8 + b$	Simplify.
$-21.8 = b$	Subtract 48.8 from both sides of the equation to solve for b.

Substituting the values we found for m and b into $y = mx + b$, we write the linear model $y = 0.8x - 21.8$. If you created a scatter plot on your calculator, you can enter this equation into Y_1 to verify that the line we found is a good model for the data. See Figure 9. This linear equation can now be used to approximate answers to various questions we might ask about the trend.

Figure 9.

▶ Example 3

Use the equation of the trend line found in Example 2 to predict the number of cricket chirps in a 15-second interval when the temperature is 78°F.

Solution

Substituting $x = 78$ into the equation $y = 0.8x - 21.8$, we obtain

$$y = 0.8(78) - 21.8$$

$$= 40.6$$

So we estimate that a cricket will chirp about 40.6 times in a 15-second interval when the outside temperature is 78°F.

If you entered the equation into your graphing calculator, you can also find this value using your [TRACE] button. See Figure 10. Make sure you are tracing on the equation in Y_1 rather than on the scatter plot. You can use up and down arrows to toggle between the data in the scatter plot and points on the line. The left and right arrows move the cursor along the curve. When tracing on the line, enter any x value in the chosen window and press [ENTER] to find the corresponding y value.

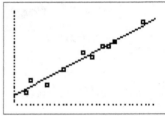
P1:L1,L2
X=78 .... Y=40.6 ....
Figure 10.

C. FINDING LINEAR REGRESSION EQUATIONS

While eyeballing a line works reasonably well, there are statistical techniques for fitting a line to data that minimize the differences between the line and data values. One such technique is called **least squares regression** and can be computed by many graphing calculators, spreadsheet software, statistical software, and many web-based calculators. Least squares regression is one way to define the line that best fits the data. We will refer to this method as **linear regression**.

Before we use our graphing calculators to find a linear regression equation, it is always a good idea to graph the scatter plot of the data first. This way we can determine if a linear model makes sense in a given situation. Besides, entering the data into Lists 1 and 2 on the calculator is a required step for finding a linear regression line.

Finding a Linear Regression Line on a Graphing Calculator

To find a linear regression on a graphing calculator, follow these steps:

Step 1: Enter the input values into L_1 and the output values in to L_2.

Step 2: From the home screen, return to [STAT] and arrow right to [CALC]. Then arrow down to select linear regression (LinReg) and press [ENTER]. See Figure 11.

Step 3: Use the displayed values of a and b to write down the linear equation,, rounding these values to a reasonable number of decimal places. Notice the calculator uses a for the slope rather than m. See Figure 12.

Step 4: Methods for viewing the equation on the graphing screen vary by calculator edition, but you can always enter the equation by hand in Y_1.

Step 5: To see both the regression equation and the scatter plot of the data, make sure Plot 1 is turned on and that you have chosen a good viewing window ([ZOOM] 9). See Figure 13.

```
EDIT CALC TESTS
1:1-Var Stats
2:2-Var Stats
3:Med-Med
4:LinReg(ax+b)
5:QuadReg
6:CubicReg
7↓QuartReg
```

Figure 11.

Figure 12.

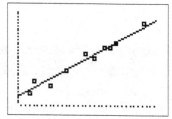

Figure 13.

▶ *Example 4*

Finding a Linear Regression Line

Find the linear regression line using the cricket-chirp data in Figure 3. Use this equation to predict the number of chirps when the temperature is 78°F.

Solution

Enter the input values (temperature) in List 1 (L_1).

Next, enter the output values (chirps) in List 2 (L_2). See Figure 14. On a graphing utility, select Linear Regression (LinReg). See Figure 15.

After rounding the a and b values, we obtain this equation:

$y = 0.828x - 23.586$

Notice that this line is quite similar to the equation we calculated by hand in Example 2, but it fits the data better. Notice also that using *this* equation will change our prediction for the number of chirps in 15 seconds at 78°F to

$$y = 0.828(78) - 23.586$$
$$= 40.998$$
$$\approx 41 \text{ chirps}$$

This prediction is just slightly greater than our prediction of 40.6 chirps in Example 2. Figure 16 shows the graph of the scatter plot with the least squares regression line.

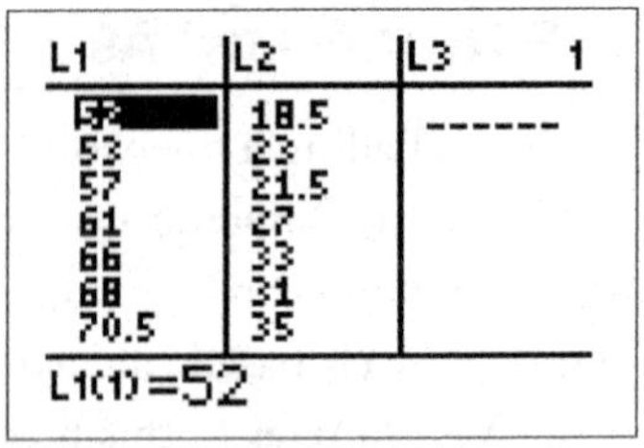

Figure 14.

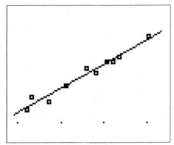

Figure 15.

Figure 16. Scatter plot with regression line.

Practice Set C

The data in the table in Figure 17 record the speed of a truck in miles per hour and its fuel efficiency in miles per gallon when traveling at that speed. Use this data to answer the questions below. Check your solutions on the following page.

1. Use a graphing calculator to draw a scatter plot of the data. Describe the trend of these points.

2. Use the calculator to find a linear regression equation to model the data. Round a and b to three decimal places.

3. Use the regression equation to estimate the fuel efficiency when the speed of the truck is 55 mph.

x mph	y mpg
20	38
33	35
35	34
40	33
48	32
60	25
65	22

Figure 17.

D. USING A LINEAR MODEL TO MAKES ESTIMATES AND PREDICTIONS

When working with a linear model, we must often evaluate the linear model at a given input. We did this in Examples 3 and 4 when we predicted the rate of cricket chirps at 78°F. We used our equation to calculate that the output would be approximately 41 chirps in a 15-second interval.

While the data for most examples does not fall perfectly on the line, the equation is our best guess as to how the relationship will behave outside of the values for which we have data. We use a process known as **interpolation** when we predict a value *inside* the domain and range of the data. We use the process of **extrapolation** when we predict a value *outside* the domain and range of the data.

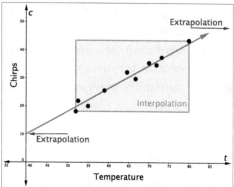

Figure 18 compares the two processes for the cricket-chirp data addressed in previous examples. We can see that interpolation will occur if we used our model to predict chirps when the values for temperature are between 52°F and 80.5°F. Extrapolation will occur if we used our model to predict chirps when the values for temperature are less than 52°F or greater than 80.5°F.

Figure 18. Interpolation occurs within the domain and range of the provided data. Extrapolation occurs outside the domain and range.

▶ *Example 5*

Understanding Interpolation and Extrapolation

Use the cricket data from Figure 3 (shown again in figure 19 for reference) and our regression model, $y = 0.828x - 23.6$, to answer the following questions:

Temperature (°F)	52	53	57	61	66	68	70.5	72	73.5	80.5
Chirps (in 15-second interval)	18.5	23	21.5	27	33	31	35	35	37	44

Figure 19. Air Temperatures vs. Cricket Chirps.

a.   Will predicting the number of chirps crickets will make during a 15-second interval when the outside temperature is 55°F be interpolation or extrapolation? Make the prediction, and discuss whether it is reasonable.

b.   Will predicting the rate of cricket-chirps at 40°F be interpolation or extrapolation? Make the prediction, and discuss whether it is reasonable.

Solution

a. The temperatures in the data are between 52°F and 80.5°F. A prediction at 55°F is inside the domain of our data, so it will be interpolation. Using our model, $y = 0.828(55) - 23.586 = 21.954 \approx 22$. We predict that crickets will chirp 22 times in a 15-minute interval. Based on the data we have, this seems reasonable.

b. Predicting the number of chirps at 40°F is extrapolation because 40 is outside the domain of our data. Using our model, $y = 0.828(40) - 23.586 = 9.534 \approx 9.5$. Our model predicts the crickets would chirp 9.5 times in 15 seconds.

While this might be possible, we have no reason to believe our model is valid outside the domain and range. In fact, it's commonly known that crickets generally stop chirping altogether below around 50 degrees.

Another way to use a linear model is to set the equation of the model equal to a specified output. If we then solve the equation, we can answer a question that asks us to find an input that will predict the specified output. For example, if we are outside on a warm evening and count 25 chirps in 15 seconds, we can use our regression model to predict the temperature.

$$y = 0.828x - 23.586$$

$$25 = 0.828x - 23.586 \qquad \text{25 chirps is an output, so } y = 25$$

$$\underline{+\ 23.586 \qquad +\ 23.586} \qquad \text{To solve for } x, \text{ add 23.586 to both sides.}$$

$$\underline{48.586 = 0.828x} \qquad \text{Simplify.}$$

$$0.828 \qquad 0.828 \qquad \text{Divide both sides by 0.828.}$$

$$58.7 \approx x$$

A count of 25 chirps in 15 seconds predicts that the temperature is approximately 58.7°F. This is an example of interpolation because the number of chirps is within the range of our data.

▶ Example 6

Using a Regression Line to Make Predictions

Gasoline consumption in the U. S. has been steadily increasing. See Figure 20 for data from 1994 to 2004.

Year	'94	'95	'96	'97	'98	'99	'00	'01	'02	'03	'04
Consumption (billions of gallons)	113	116	118	119	123	125	126	128	131	133	136

Figure 20.

Practice Set C — Answers

1. The data appear to be approximately linearly related. As the speed of the truck increases the fuel efficiency decreases. This indicates a linear function with a negative slope could model the data.

2. $y = -0.351x + 46.364$

3. $y = -0.351(55) + 46.364 \approx$ 27.1 At a speed of 55mph, the fuel efficiency of the truck will be approximately 27 miles per gallon.

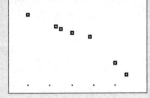

Using the data from Figure 20, do the following.

    a. Determine whether the trend is linear, and if so, find a model for the data.

    b. Interpret the slope and y-intercept of the model in the context of the problem.

    c. Use the model to predict the consumption in 2008.

    d. Use the model to predict when the consumption will reach 175 billion gallons.

Solution

In a problem that uses years like this, it's useful to define the independent variable with respect to a starting year. This ensures that the input values will be smaller, more manageable numbers. So, let x be the number of years since 1994. Then $x = 0$ corresponds to 1994, $x = 1$ corresponds to 1995, and so on. Let y be the gas consumption in the U.S. in billions of gallons.

    We can use the table in Figure 21 for our input-output pairs in List 1 and List 2 on the graphing calculator.

x Years since 1994	0	1	2	3	4	5	6	7	8	9	10
y Consumption (billion of gallons)	113	116	118	119	123	125	126	128	131	133	136

Figure 21.

a. We create a scatter plot using a calculator and see that the data are nearly linear with a positive slope. We then calculate the linear regression line and graph it along with our scatter plot as shown in Figure 22. By rounding the values of a and b, our regression equation is $y = 2.209x + 113.318$.

    Figure 22 presents the scatter plot of the data, including the regression line.

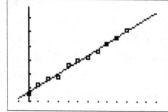

Figure 22. Scatter plot of data years vs. consumption.

b. The slope indicates that gasoline consumption in the U.S. has been increasing at an average rate of 2.209 billion gallons per year. The y-intercept indicates that gas consumption in the initial year, 1994 for our data set, was approximately 113.3 billion gallons.

c. For the year 2008, $x = 2008 - 1994 = 14$. We calculate that $y = 2.209(14) + 113.318 = 144.244$.

    The model thus predicts 144.244 billion gallons of gasoline consumption in 2008. This prediction is an extrapolation since the year 2008 is outside of the domain of the data set.

d. To predict when the consumption will reach 175 billion gallons, substitute 175 for y in our equation and solve for x.

$$175 = 2.209x + 113.318$$

$$61.682 = 2.209x \qquad \text{Subtract 113.318 from both sides and simplify.}$$

$$27.923 \approx x \qquad \text{Divide both sides by 2.209 and simplify.}$$

Our solution $x \approx 27.923 \approx 28$ years corresponds to the year 1994 + 28 — 2022. So we predict that gas consumption in the U.S. will reach 175 billion gallons in 2022. This prediction is also an extrapolation.

Practice Set D

Use the model we created in Example 6 to answer the questions below. Turn the page to check your solutions.

4. Predict the gas consumption in 2017. Is this an interpolation or an extrapolation?

5. Predict the year the gas consumption will reach 200 billion gallons, assuming the model is still relevant that far into the future.

E. INTERCEPTS OF A MODEL AND MODEL BREAKDOWN

There's a big difference between making predictions inside the domain and range of values for which we have data and making predictions outside that domain and range. Predicting a value outside the domain and range has its limitations.

In Example 6, the U.S. gas consumption problem, the further we move beyond the domain of the provided data — that is, the farther into the future we predict — the less sure we can be about our predictions. The rate of gas consumption *could* increase or decrease dramatically, but it might also remain constant because of the increase of electric or hybrid vehicles, the use of alternative energy, or other changes that would affect gasoline consumption. When we extrapolate much beyond our data set, we lose confidence in our prediction.

When a model no longer applies beyond a certain point because a value of the input or output variable is obviously wrong based on common knowledge it is called **model breakdown**.

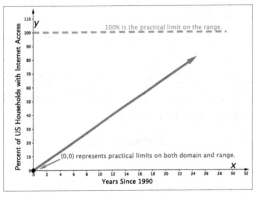

For example, consider a linear model that represents y as the percentage of U.S. homes with Internet access x years since 1990. Clearly that percentage has been increasing over the years. However, the input values (years) are limited by the fact that *no one* had Internet access from their home before 1990. The y-intercept, when $x = 0$, indicates a practical limit on our domain. The x-intercept, when $y = 0$, indicates that 0 percent of homes have Internet access. The output values (percentages of homes) can only be between 0 and 100%, representing a practical limit on our range. See Figure 23.

Figure 23.

▶ *Example 7*

Students in a college dormitory carried out an experiment on hours of TV watched per day (x) and number of sit-ups a person can do (y). After collecting data and plotting a scattergram, the students could see that there was a linear relationship between the two activities. They then ran a linear regression resulting in the equation $y = -1.341x + 32.234$.

    a. Determine the y-intercept of the regression model and interpret its meaning in the context of the problem.

    b. Use the regression model to find the x- intercept. Interpret its meaning in the context of the problem. Discuss the potential for model breakdown.

    c. Determine a practical domain for this function. Justify your answer.

Solution

a. The y-intercept is (0, 32.234). It means that a student in the dorm experiment who watched no TV could do about 32 sit-ups.

b. Setting the equation equal to 0 and solving for x we have

$$0 = -1.341x + 32.234$$
$$-32.234 = -1.341x$$
$$24.037 \approx x$$

The x-intercept is thus approximately (24, 0), which means that student who watched TV for 24 hours per day would not be able to do any sit-ups. This situation represents model breakdown because it's probable that no one can watch TV for 24 hours a day, however hard they might try.

c. Determining the practical domain is partly subjective and correct answers might vary. One reasonable practical domain could be [0, 16], representing from 0 to 16 hours of TV time. It's impossible to have a negative number of hours watching TV, and it is unlikely that students would watch TV for more than 16 hours per day. They do have classes to attend.

Exercises

1. What is interpolation when using a linear model?

2. What is extrapolation when using a linear model?

3. Explain the difference between a positive and a negative linear relationship.

4. What does it mean to say that model breakdown has occurred?

For the following exercises, draw a scatter plot for the data provided. Does the data appear to be linearly related? If so, is it a positive or negative linear relationship?

5.

x	0	1	3	4	7	8	10
y	22	23	19	15	11	6	5

6.

x	1	2	3	4	5	6	7
y	46	50	59	75	100	136	185

7.

x	100	190	250	310	380	430	450
y	12	28	25	43	60	73	75

8.

x	1	3	5	6	7	9	11
y	1	9	28	57	65	125	216

9. The town of Midgar increased in population from 1990 to 2010. For the following data, draw a scatter plot and estimate a line of best fit. Choose two points on or near your line and use them to write a linear model for the data. Let y be the population x years since 1990. Use your model to predict the population in 2018. Does this prediction involve interpolation or extrapolation?

Year	1990	1995	2000	2005	2010
Population	11,500	12,100	12,700	13,000	13,750

10. Vincent is defrosting frozen vegetables in the microwave. For the following data, draw a scatter plot and estimate a line of best fit. Choose two points on or near your line and use them to write a linear model for the data. Use your model to predict when the temperature would reach 48°F. Does this prediction involve interpolation or extrapolation?

Time, seconds	46	50	54	58	62
Temperature, °F	26	29	31	35	40

For the following exercises, match each scatterplot with one of the four specified relationships in Figure 24.

11. Nearly linear with a positive slope

12. Nearly linear with a negative slope

13. No relationship

14. Far from linear with a possible negative slope.

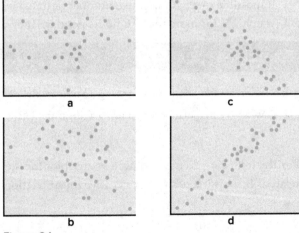

Figure 24.

15. A regression was run to determine whether there is a relationship between years since 1990 (x) and number of people in millions in the U.S. labor force (y). The results of the regression are shown below. Assuming the trend continues, use this to predict the number of people in the labor force in 2012 and in 2020.

$$y = ax + b$$
$$a = 1.7$$
$$b = 124$$

16. A regression was run to determine whether there is a relationship between the diameter of a tree (x, in inches) and the tree's age (y, in years). The results of the regression are shown below. Use this to predict the age of a tree with diameter 10 inches and a tree with diameter 15 inches.

$$y = ax + b$$
$$a = 6.301$$
$$b = -1.044$$

17. The U.S. Census tracks the percentage of persons 25 years or older who are college graduates. Figure 25 provides several years worth of data. Use your graphing calculator to make a scatter plot of the data and to find a linear regression equation to model the data. Assuming the trend continues, in what year will the percentage exceed 35%?

Year	1990	1992	1994	1996	1998	2000	2002	2004	2006	2008
Graduates (percent)	21.3	21.4	22.2	23.6	24.4	25.6	26.7	27.7	28	29.4

Figure 25.

18. Figure 26 provides the U.S. import of wine (in hectoliters) for several years. Let y be the number of hectoliters import x years after 1990. Find a linear regression equation to model the data. Assuming the trend continues, use your model to predict when imports will exceed 12,000 hectoliters.

Year	1992	1994	1996	1998	2000	2002	2004	2006	2008	2009
Imports	2665	2688	3565	4129	4584	5655	6549	7950	8487	9462

Figure 26.

19. Figure 27 shows the year and the number of people unemployed in the City of Waldorf over several years. Let y be the number of people who are unemployed x years after 1990. Find a linear regression equation and, assuming the trend continues, use it to predict the number unemployed in 2016.

Year	1990	1992	1994	1996	1998	2000	2002	2004	2006	2008
Number Unemployed	750	670	650	605	550	510	460	420	380	320

Figure 27.

Practice Set D — Answers

4. According to the model, gas consumption in 2017 ($x = 23$) is 164.125 billion gallons. This number was found through extrapolation.

5. According to the model, gas consumption will reach 200 billion gallons when $x \approx 39.240 \approx 39$. This corresponds to the year 2033.

20. Figure 28 shows a small sample of heights and weights of women aged 30 – 39. Find a linear regression equation to model this data. Let y be the weight of a woman who is x inches tall. Use your model to predict the height of a 120-pound woman.

Height (inches)	62	67	68	70	71	74	75
Weight (pounds)	123.1	138.1	134.3	151	145.5	164.2	176.4

Figure 28.

21. The Oregon Department of Education is conducting a review of mathematics education over the last 25 years. Using Figure 29, determine if there is a linear trend between the number of years since 1990 (x) and percentage of students who took Algebra 2 (y). Find a linear regression equation to model this data and use it to predict the percentage of high school students taking Algebra 2 in 2020.

x (year)	1992	1997	2000	2004	2008	2010	2015
y (Percentage of high school students)	39.9	49.0	58.8	61.5	61.7	67.6	70.3

Figure 29.

22. Veronica is looking through records in her hometown of Silver Creek. She is comparing the average age of mothers at the time of the birth of their first child (y) with the year that the births took place. Let x be the number of years since 1980. Using the data from figure 30, determine if there is a linear trend, and, if there is, find a linear regression equation and use it to predict the average age of a first-time mother in the year 2020.

Year	1980	1983	1985	1988	1990	1993	1995	1998	2000	2003
Age of mother	19	22	21	23	23	24	22	23	21	25

Figure 30.

23. The data in Figure 31 shows both the number of calories and the number of grams of fat in seven sandwiches at a McDonald's restaurant. Find a linear regression equation to model the data and use it to predict the number of grams of fat in a sandwich with 620 calories.

Sandwich Name	Calories	Fat (grams)
Big Mac	550	29
Quarter Pounder w/ cheese	520	26
Double quarter-pounder w/ cheese	750	42
Hamburger	250	9
Hamburger w/ cheese	300	12
Double Cheeseburger	440	23
McDouble	390	19

Figure 31.

1.5 Function Notation and Making Predictions

OVERVIEW

Once we determine that a relationship is a function, we want to display and define the function so that we can easily understand and use it. As we have already seen, there are various ways of representing functions. Standard function notation is one representation that makes working with functions easier. In this section, we introduce a new notation for functions commonly used by mathematicians, scientists, and computer programmers. You will learn to:

- Use function notation to represent functions.

- Find inputs and solve for outputs using function notation.

- Use function notation with graphs and tables.

- Interpret models described with function notation.

- Make predictions from models described with function notation.

- Find and interpret intercepts, domain, and range of a model.

A. FUNCTION NOTATION

When using an equation, graph, table, or words to represent a function, it is very useful to name the function. The letters f, g, and h are often used to name functions although any letter can be used.

Once we name a function, say f, and if we are using x and y as input and output variables, we can use the symbol $f(x)$ to represent the output y. We refer to "$f(x)$" as **function notation** and we read the expression as "f of x". We use parentheses to contain the function input, in this case x. Note in this context the parentheses *do not indicate multiplication*.

> ### Function Notation
>
> The notation $y = f(x)$ defines a function named f. The letter x represents the input value, or independent variable. The letter y, or $f(x)$, represents the output value, or dependent variable.

Since $f(x)$ is another name for the output, y, we can write $y = f(x)$ which nicely communicates that y is a function of x.

When we know an input value and want to determine the corresponding output value for a function, we **evaluate** the function. Evaluating will always produce one result because each input value of a function corresponds to exactly one output value.

When we have a function in algebraic form, it's usually a simple matter to evaluate the function. We replace the input variable with a given value and then simplify according to the correct order of operations (PEMDAS).

For example, we can evaluate the function $f(x) = 5 - 3x^2$ by squaring the input value, multiplying by 3, and then subtracting the product from 5. To evaluate this function when $x = 2$, we write:

$$
\begin{aligned}
f(x) &= 5 - 3x^2 && \text{Equation of } f. \\
f(2) &= 5 - 3(2)^2 && \text{Replace } x \text{ on both sides with the input value 2.} \\
&= 5 - 3(4) && \text{Simplify according to PEMDAS.} \\
&= 5 - 12 \\
&= -7
\end{aligned}
$$

The equation $f(2) = -7$ means the input $x = 2$ leads to the output $y = -7$. Notice this equation is of the form

$$f(\text{input}) = \text{output}$$

This is the essential meaning of function notation.

▶ Example 1

Evaluating Functions at Specific Values

For the functions $f(x) = 6x - 5$ and $g(x) = x^3 + 1$, evaluate the functions at $x = 3$, $x = -4$, and $x = 2.59$.

Solution

First, we replace the x in the functions with each specified value, starting with f and $x = 3$:

$$
\begin{aligned}
f(3) &= 6(3) - 5 && \text{Substitute 3 for } x. \\
&= 18 - 5 && \text{Multiply.} \\
&= 6 && \text{Subtract.}
\end{aligned}
$$

Now with $x = -4$:

$$
\begin{aligned}
f(-4) &= 6(-4) - 5 && \text{Substitute } -4 \text{ for } x. \\
&= -29 && \text{Simplify.}
\end{aligned}
$$

And $x = 2.59$:

$$
\begin{aligned}
f(2.59) &= 6(2.59) - 5 && \text{Substitute 2.59 for } x. \\
&= 10.54 && \text{Simplify.}
\end{aligned}
$$

Next, we replace the x in g with $x = 3$:

$$
\begin{aligned}
g(3) &= (3)^3 + 1 && \text{Substitute 3 for } x. \\
&= 27 + 1 && \text{Perform exponentiation (raise 3 to the 3}^{\text{rd}}\text{ power).} \\
&= 28 && \text{Add.}
\end{aligned}
$$

Now with $x = -4$:

$$g(-4) \;=\; (-4)^3 + 1 \qquad \text{Substitute } -4 \text{ for } x.$$
$$ \;=\; -63 \qquad\qquad \text{Simplify.}$$

And $x = 2.59$:

$$g(2.59) \;=\; (2.59)^3 + 1 \qquad \text{Substitute } 2.59 \text{ for } x.$$
$$ \;\approx\; 18.374 \qquad\qquad \text{Simplify.}$$

We define some functions with the variable in more than one place, for example $f(x) = x^2 + 3x - 4$. In this case, when we evaluate the function for a specific input value, we must substitute that value everywhere we find an x.

We can also give an algebraic expression as the input to a function. For example, $f(a + 5)$ means, "Evaluate the function at $a + 5$."

▶ *Example 2*

Evaluating a Function

For $f(x) = x^2 + 3x - 4$ find the following:

   a. $f(-2)$

   b. $f(a)$

   c. $f(a + 5)$

Solution

a. Because the input value is a number, -2, we just use the correct order of operations to simplify.

$$f(-2) \;=\; (-2)^2 + 3(-2) - 4 \qquad \text{Substitute } -2 \text{ for } x.$$
$$ \;=\; -6$$

b. In this case, the input value is a letter, and we cannot simplify the answer.

$$f(a) \;=\; (a)^2 + 3(a) - 4 \qquad \text{Substitute } a \text{ for } x.$$
$$ \;=\; a^2 + 3a - 4$$

c. With an input value of $a + 5$, we must use the **distributive property** to simplify.

$$f(a + 5) \;=\; (a + 5)^2 + 3(a + 5) - 4 \qquad\quad \text{Substitute } a + 5 \text{ for } x.$$
$$ \;=\; (a + 5)(a + 5) + 3(a + 5) - 4 \quad \text{Expand.}$$
$$ \;=\; a^2 + 10a + 25 + 3a + 15 - 4 \quad \text{Use the distributive property and FOIL.}$$
$$ \;=\; a^2 + 13a + 36 \qquad\qquad\qquad\; \text{Combine like terms.}$$

In the previous example we see for an algebraic function when an input is a number, the output will be a number. However, when the input is a variable expression, the output will be a variable expression.

Sometimes we know an output value and want to determine the input value(s) that would produce that output value. In these cases we set the output equal to the function's formula and solve for the input.

▶ Example 3

Using an Equation to Find Input and Output

Given the function $f(x) = \frac{1}{2}x + 9$

    a. Find $f(10)$.

    b. Find x when $f(x) = -6$.

Solution

a.

$$f(10) = \tfrac{1}{2}(10) + 9 \qquad \text{Substitute 10 for } x.$$
$$= 14 \qquad\qquad \text{Simplify.}$$

b. We substitute -6 for $f(x)$ and solve for x. Notice that we set the output value on the right side of the equation, but it can be set on the left side as well.

$$\tfrac{1}{2}x + 9 = -6 \qquad \text{Substitute } -6 \text{ for } f(x).$$
$$\tfrac{1}{2}x = -15 \qquad \text{Subtract 9 from both sides.}$$
$$x = -30 \qquad \text{Multiply both sides by 2 (the reciprocal of } \tfrac{1}{2}\text{).}$$

When solving for an input value, given an output value, we sometimes get more than one solution. This is because for some functions different input values can produce the same output value.

▶ Example 4

Using an Equation to Find Input and Output

Given the function $f(x) = x^2 - 3$

    a. Find $f(5)$ and $f(-5)$.

    b. Solve $f(x) = 97$.

Solution

a. Notice that for this function, the two given inputs produce the same output.

$$f(5) = (5)^2 - 3 \qquad \text{Substitute 5 for } x.$$
$$= 22 \qquad\qquad \text{Simplify.}$$

And:

$$f(-5) = (-5)^2 - 3 \qquad \text{Substitute } -5 \text{ for } x.$$
$$= 22 \qquad \text{Simplify.}$$

b. We substitute 97 for $f(x)$ and solve for x.

$$x^2 - 3 = 97 \qquad \text{Substitute 97 for } f(x).$$
$$x^2 = 100 \qquad \text{Add 3 to both sides.}$$
$$x = \pm 10 \qquad 10^2 = 100 \text{ and } (-10)^2 = 100$$

Both inputs, $x = 10$ and $x = -10$, produce the same output: $f(x) = 97$.

Practice Set A

When you're finished with these problems, turn the page and check your solutions.

1. Given the function $f(x) = \sqrt{x - 4}$, evaluate the function at $x = 40$ and $x = 15$.

2. Given the function $g(x) = 5x^2 + x$, find $g(3.4)$, $g(-2)$, and $g(a - 1)$.

3. Given the function $h(x) = 8 - 2.5x$, find x when $h(x) = -10.5$.

B. USING FUNCTION NOTATION WITH GRAPHS AND TABLES

Functions in Tabular Form

As we saw in Section 1.2, we can represent functions in tables. Let's now apply function notation to the process of determining input and output values *from* a table. When a function is represented by a table, the process for evaluating the function at a given input value is as follows

1. Find the given input in the row (or column) of input values.

2. Identify the corresponding output value paired with that input value.

To determine the input or inputs that resulted in a given output, do the following:

1. Find the given output values in the row (or column) of output values.

2. Identify the input value(s) corresponding to the given output value.

▶ *Example 5*

Finding Function Values from a Table

Use the table in Figure 1 to find the values asked for.

a. Find $f(2)$.

b. Find $f(17)$.

c. Find t when $f(t) = 26$.

d. Find t when $f(t) = 11$.

t	$f(t)$
2	11
5	13
9	17
12	26
17	18
21	15
25	11

Figure 1.

Solution

a. The value in parentheses always represents the input value, in this case $t = 2$. The corresponding output value in our table is 11. So $f(2) = 11$.

b. The number 17 occurs in both the input and output columns, so be careful here! In this case, the input value $t = 17$. The corresponding output value in our table is 18. So $f(17) = 18$.

c. This time we are given the *output* value and asked to solve for the input value. We find 26 in the second column of the table. The corresponding input value is 12. So $t = 12$.

d. Again, we are given the *output* value. Notice that the number 11 occurs *twice* in the output column. There are two corresponding inputs, 2 and 25. So $t = 2$ and $t = 25$.

Finding Function Values from a Graph

For functions represented by a graph, the function notation $y = f(x)$ means that the point (x, y) is on the graph of f. For example, if for some function f, $f(7) = 11$, then the point $(7, 11)$ is on its graph.

Evaluating a function requires finding the corresponding output value for a given input value. When the function is represented by a graph, we first locate the point on the graph that has the given input value. We then read off the y-coordinate of the point. Finding the input value of a function equation using a graph requires finding all instances of the given output value on the graph and observing the corresponding input value(s).

Practice Set A — Answers

1. $f(40) = \sqrt{40 - 4} = \sqrt{36} = 6$
$f(15) = \sqrt{15 - 4} = \sqrt{11} \approx 3.317$

2. $g(3.4) = 5(3.4)^2 + (3.4) = 61.2$
$g(-2) = 5(-2)^2 + (-2) = 5(4) - 2 = 18$

$$
\begin{aligned}
g(a - 1) &= 5(a - 1)^2 + (a - 1) &&\text{Substitute } a - 1 \text{ for } x. \\
&= 5(a - 1)(a - 1) + a - 1 &&\text{Expand.} \\
&= 5(a^2 - 2a + 1) + a - 1 &&\text{Expand further using the FOIL method.} \\
&= 5a^2 - 10a + 5 + a - 1 &&\text{Multiply by 5 according to the distributive property.} \\
&= 5a^2 - 9a + 4 &&\text{Combine like terms.}
\end{aligned}
$$

3. In the equation, replace $g(x)$ with -10.5 and solve for x.

$$
\begin{aligned}
8 - 2.5x &= -10.5 &&\text{Substitute } -10.5 \text{ for } g(x). \\
-2.5x &= -18.5 &&\text{Subtract 8 from both sides.} \\
x &= 7.4 &&\text{Divide both sides by } -2.5.
\end{aligned}
$$

▶ *Example 6*

Reading Function Values from a Graph

Given the graph in Figure 2,

    a. Evaluate $f(2)$.   **b.** Find x when $f(x) = 4$.

Solution

a. To evaluate $f(2)$, locate the point on the curve where $x = 2$, then read the y-coordinate of that point. The point has coordinates $(2, 1)$, so $f(2) = 1$. See Figure 3.

b. To solve $f(x) = 4$, find the output value 4 on the vertical axis. Moving horizontally along the line $y = 4$, locate two points of the curve with output value 4: $(-1, 4)$ and $(3, 4)$. So the two solutions are $x = -1$ or $x = 3$. See Figure 4.

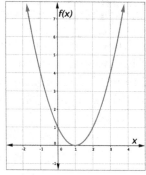

Figure 2.

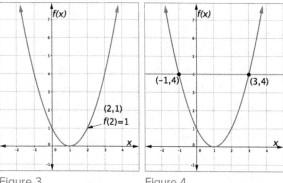

Figure 3.   Figure 4.

Practice Set B

Now it's your turn. When you are finished, turn the page and check your solutions.

4. Use the table to find the values of function h.

t	0	5	10	15	20	25	30
$h(t)$	0	12	15	20	25	12	3

    a. Find $h(15)$.

    b. Find $h(30)$.

    c. Find t when $h(t) = 20$.

    d. Find t when $h(t) = 12$.

5. Use the graph to find the values asked for.

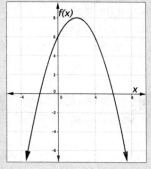

    a. Find $f(0)$.

    b. Find $f(5)$.

    c. Find x when $f(x) = 8$.

    d. Find x when $f(x) = 0$.

C. MAKING PREDICTIONS FROM MODELS DESCRIBED WITH FUNCTION NOTATION

So far, we've mostly used our favorite variables, x and y, to name the input-output pairs of a function. When working with a real world problem, it's common to choose letters that will help us remember what the variables represent in the problem. For example, to represent a relation where "height is a function of age," we start by identifying the descriptive variables h for height and a for age.

h is f of a We name the function f. Height is a function of age.

$h = f(a)$ We use parentheses to indicate the function input.

$f(a)$ The expression is read as "f of a."

The notation $h = f(a)$ shows us that h depends on a. The value a must be put into the function h to get a result. Remember, the parentheses indicate that age is an input to the function — they do *not* indicate multiplication.

Notice that the inputs to a function do not have to be numbers. Function inputs can be names of people, labels of geometric objects, or any other element that determines some kind of output. However, most of the functions that we will work with in this book will have numbers as inputs and outputs.

▶ *Example 7*

Using Function Notation for Days in a Month

Use function notation to represent a function whose input is the name of a month and whose output is the number of days in that month.

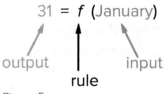

Figure 5.

Solution

The number of days in a month is a function of the name of the month, so if we name the function f, we write days = f(month) or d = $f(m)$. The name of the month is the input to a "rule" that associates a specific number (the output) with each input. For example, f(March) = 31 because March has 31 days and f(June) = 30 because June has 30 days. The notation $d = f(m)$ reminds us that the number of days, d (the output), is dependent on the name of the month, m (the input).

Practice Set B — Answers

4.

  a. $h(15) = 20$

  b. $h(30) = 3$

  c. $t = 15$

  d. $t = 5$ and $t = 25$

5.

  a. $f(0) = 6$

  b. $f(5) = 3.5$ (This is a bit of a guess as the curve does not fall exactly on the intersection of gridlines when $x = 5$.)

  c. $x = 2$

  d. $x = -2$ and $x = 6$

▶ *Example 8*

Interpreting Function Notation

A function $N = f(y)$ gives the number of police officers, N, in Salem in year y. What does $f(2016) = 300$ represent?

Solution

When we read $f(2016) = 300$, we see that the input year is 2016. The value for the output, the number of police officers N, is 300. Remember, $N = f(y)$. The statement $f(2016) = 300$ tells us that in the year 2016, there were 300 police officers in Salem.

Now let's combine the skills we've learned in this section and apply them to the following problem.

▶ *Example 9*

Interpreting Function Notation and Using Function Notation to Make Predictions

The average cost of in-state tuition and fees at public, four-year universities has been increasing approximately linearly over the last two decades. A function that models this situation is given by $C = f(t) = 320t + 3030$, where C is the cost at t years since the year 2000.

   a. Find $f(0)$ and interpret its meaning in the context of the problem.

   b. Find $f(20)$ and interpret its meaning in the context of the problem.

   c. Find t when $f(t) = 11,350$ and interpret the solution in the context of the problem.

Solution

a. $f(0) = 320(0) + 3030 = 3030$
The input value 0 represents years since 2000, while the output value 3030 represents cost. So our interpretation is that in the year 2000, the average cost of in-state tuition and fees at public, four-year universities was $3030.

b. $f(20) = 320(20) + 3030 = 9430$
We predict that in the year 2020, the average cost of tuition and fees at community colleges will be $9430.

c. Since we are given an output value and asked to find an input value, substituute 11,350 for $f(t)$ and solve for t.

$$320t + 3030 \;=\; 11{,}350 \qquad \text{Substitute 11,350 for } f(t).$$
$$320t \;=\; 8320 \qquad \text{Subtract 3030 from both sides.}$$
$$t \;=\; 26 \qquad \text{Divide both sides by 320.}$$

We predict the cost of in-state tuition and fees at public, four-year universities will be $11,350 in the year 2026.

Practice Set C

Now it's your turn to use function notation with the following problems. You can turn the page to check your solutions.

6. Use function notation to express the weight of a pig in pounds (w) as a function of its age in days (d).

7. The function $L = f(g)$ gives the length of a hanging spring, L, in centimeters when a weight of g grams is added to the spring. Interpret $f(100) = 15$.

8. The pressure, P, in pounds per square inch (psi) on a scuba diver depends upon her depth below the water surface, d, in feet. This relationship may be modeled by the equation,
$P = f(d) = 0.434d + 14.696$.

  a. Find and interpret $f(20)$.

  b. Find d when $f(d) = 35$.

D. FINDING INTERCEPTS OF A MODEL USING FUNCTION NOTATION

Let's look at vertical and horizontal intercepts of a graph of a function when its equation is given in function notation — this time with variables other than our favorite variables x and y.

For example, if a function were described by $q = f(n)$, then the n-intercept is the point $(n, 0)$ where the function intersects the horizontal axis. This is because the input or independent variable is always graphed on the horizontal axis. The q-intercept is the point $(0, q)$ where the function intersects the vertical axis.

It can help to sketch a graph of the function to see how this works. Figure 6 is a possible graph of $q = f(n)$.

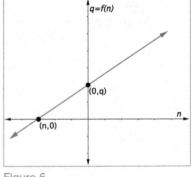

Figure 6.

▶ Example 10

Finding and Interpreting Intercepts of a Model Described with Function Notation

The U.S. Market share (percentage of U.S. vehicle sales) of Ford vehicles has been decreasing in recent years. Suppose the function $M = f(t) = -1.2t + 35.3$ models this situation where M is the market share in percents at t years since 1990.

  a. Find the M-intercept. Interpret its meaning in the context of the problem.

  b. Find and interpret $M = f(21)$.

  c. Find the t-intercept. Interpret its meaning in the context of the problem.

Solution

a. The *M*-intercept is the vertical intercept and occurs when $t = 0$.

$$f(0) = -1.2(0) + 35.3 = 35.3$$

This means that in 1990, Ford had a U.S. market share of 35.3%. In other words, 35.3% of the vehicles sold in the U.S. in 1990 were Fords.

b. $f(21) = -1.2(21) + 35.3 = 10.1$

In 2011 — 21 years after 1990 — Ford's U.S. market share was only 10.1%.

c. The *t*-intercept is the horizontal intercept and occurs when $m = 0$. In the equation replace $f(t)$ with 0 and solve for *t*.

$$-1.2t + 35.3 = 0 \qquad \text{Substitute 0 for } f(t).$$
$$-1.2t = -35.3 \qquad \text{Subtract 35.3 from both sides.}$$
$$t \approx 29.4 \qquad \text{Divide both sides by } -1.2.$$

We predict that in 2019, 29 years after 1990, Ford's market share will decrease to 0%. However, this solution is most likely an example of **model breakdown**. In all likelihood, Ford will still be making and selling vehicles in the U.S. in 2019

A graphical representation of a function helps us visualize the relationship between the variables, and that visualization helps us interpret applications.

Figure 7 gives us the graph of the linear function representing Ford's market share over time. The solutions to Example 9 are labeled on the graph. The *m*-intercept shows that in 1990 ($t = 0$) Ford's market share was 35.3%. By the year 2011, it had decreased to 10.1%. If the trend continues, it looks like Ford's market share will be zero in 2019, which is likely a breakdown of the applicability of this linear model.

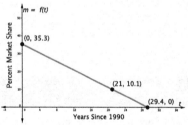

Figure 7. The graph of $y = -1.2x + 35.3$.

Practice Set D

Eduardo and his best friend are planning a long back-packing trip. They plan to hike a stretch of the Pacific Crest Trail (PCT) with a pace of about 15 miles per day. The function $R = f(d) = -15d + 212$ models the number of miles *remaining* after *d* days on the trail.

9. Find and interpret $f(6)$.

10. Find and interpret the *R*-intercept.

11. Find and interpret the *d*-intercept.

Turn the page to check your solutions.

Practice Set C — Answers

6. $w = f(d)$

7. When a 100 gram weight is added to the spring, the length of the spring will be 15 centimeters.

8. For this function

a. $P = f(20) = 0.434(20) + 14.696 = 23.376$

When the diver is 20 feet below the water surface there are 23.376 psi of pressure on the diver.

b. In the equation replace $f(d)$ with 35 and solve for d.

$0.434d + 14.696 = 35$	Substitute 35 for $f(d)$.
$0.434d = 20.304$	Subtract 14.696 from both sides.
$t \approx 46.783$	Divide both sides by 0.434.

We predict the pressure on the diver will be 35 psi when she is approximately 46.783 feet below the surface.

Practice Set D — Answers

9. $f(6) = -15(6) + 212 = 122$
This means that after 6 days, Eduardo and his friend will have 122 miles remaining for their trip.

10. The R-intercept, or vertical intercept occurs where $d = 0$.
$f(0) = -15(0) + 212 = 212$

This means that at the beginning of the hike, they have 212 miles ahead of them.

11. Replace R with 0 and solve for d.
$-15d + 212 = 0 \rightarrow d = 14.333$. This means that after a little more than 14 days, Eduardo and his friend will be done with the hike. There will be zero miles remaining.

Exercises

For the following exercises, evaluate the function f at the indicated values $f(-3)$, $f(2)$, and $f(a + 3)$.

1. $f(x) = 2x - 5$

2. $f(x) = \sqrt{2 - x} + 5$

3. $f(x) = 3x^2 + 2x - 1$

4. $f(x) = \dfrac{6x - 1}{5x + 2}$

5. Given the function $k(t) = 2t - 1$:

    a. Evaluate $k(2)$.

    b. Find t when $k(t) = 7$.

6. Given the function $f(x) = 8 - 3x$:

    a. Evaluate $f(-2)$.

    b. Find x when $f(x) = -1$.

7. Given the function $p(c) = c^2 + 6$:

    a. Evaluate $p(-3)$.

    b. Find c when $p(c) = 31$.

8. Given the function $f(x) = 3 - 3x^2$:

    a. Evaluate $f(5)$.

    b. Find x when $f(x) = 0$.

9. Given the function $f(x) = \sqrt{x} + 2$

    a. Evaluate $f(9)$

    b. Find x when $f(x) = 4$

10. Given the function $r(x) = \dfrac{x^3}{2}$

    a. Evaluate $r(-1)$

    b. Find r when $r(x) = 32$

11. Given the following graph,

    a. Evaluate $f(-1)$.

    b. Find x when $f(x) = 3$.

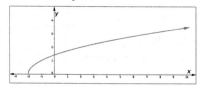

12. Given the following graph,

    a. Evaluate $f(0)$.

    b. Find x when $f(x) = -3$.

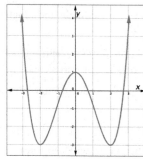

13. Given the following graph,

    a. Evaluate $f(4)$.

    b. Find x when $f(x) = 1$.

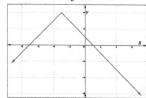

14. Given the following graph,

    a. Evaluate $f(4)$

    b. Find x when $f(x) = 8$

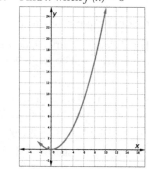

For the following exercises, use the function f represented in the following tables.

15.

x	$f(x)$
0	74
1	28
2	1
3	53
4	56
5	3
6	36
7	45
8	14
9	47

    a. Evaluate $f(3)$.

    b. Find x when $f(x) = 1$.

16.

x	$f(x)$
0	0
1	1
2	1
3	2
4	3
5	5
6	8
7	13
8	21
9	34

    a. Evaluate $f(8)$

    b. Find x when $f(x) = 1$

For the following exercises, evaluate the function f at the value $f(-2)$, $f(-1)$, $f(0)$, and $f(1)$.

17. $f(x) = 4 - 2x$

18. $f(x) = 8 - 3x$

19. $f(x) = x^2 - 7x + 3$

20. $f(x) = 3 + \sqrt{x + 3}$

21. $f(x) = \dfrac{x - 2}{x + 3}$

22. $f(x) = x^3$

23. The amount of garbage, G, produced by a city with population p is given by $G = f(p)$. G is measured in tons per week, and p is measured in thousands of people.

    a. The town of Tola has a population of 40,000 and produces 13 tons of garbage each week. Express this information in terms of the function f.

    b. Explain the meaning of the statement $f(5) = 2$.

24. The number of cubic yards of dirt, D, needed to cover a garden with area a square feet is given by $D = g(a)$.

    a. A garden with area 5000 square feet requires 50 cubic yards of dirt. Express this information in terms of the function g.

    b. Explain the meaning of the statement $g(100) = 1$.

25. Let $f(t)$ be the number of ducks in a lake t years after 1990. Explain the meaning of each statement:

    a. $f(5) = 30$

    b. $f(10) = 40$

26. Let $h(t)$ be the height above ground, in feet, of a rocket t seconds after launching. Explain the meaning of each statement:

    a. $h(1) = 200$

    b. $h(2) = 350$

27. For the following exercise, consider this scenario: The weight of a newborn, Arlo, is 7.5 pounds. The baby gained one-half pound a month for its first year.

    a. Find the linear function that models Arlo's weight W as a function of the age of the baby, in months, t.

    b. Find a reasonable domain and range for the function W.

    c. If the function W is graphed, find and interpret the t and W-intercepts.

    d. From the graph, find and interpret the slope of the function.

    e. When did the baby weigh 10.4 pounds?

    f. What is the output when the input is 6.2?

28. For the following exercise, consider this scenario: The number of math students afflicted with the common cold in the winter months steadily decreased by 205 each year from 2005 until 2010. In 2005, 12,025 people were afflicted.

    a. Find the linear function that models the number of people inflicted with the common cold C as a function of the year, t.

    b. Find a reasonable domain and range for the function C.

    c. If the function C is graphed, find and interpret the x and y-intercepts.

    d. If the function C is graphed, find and interpret the slope of the function.

    e. When will the output reach 0?

    f. In what year will the number of people be 9,700?

CHAPTER 2

Exponential Functions

Focus in on a square centimeter of your skin. Look closer. Closer still. If you could look closely enough, you would see hundreds of thousands of microscopic organisms. They are bacteria, and they're not only on your skin, but in your mouth, nose, and even your intestines. In fact, the bacterial cells in your body at any given moment outnumber your own cells. However, while some bacteria can cause illness, many are healthy and even essential to the body.

Bacteria commonly reproduce through a process called binary fission, during which one bacterial cell splits into two. When conditions are right, bacteria can reproduce very quickly. Unlike humans and other complex organisms, the time required to form a new generation of bacteria is often a matter of minutes or hours, as opposed to days or years.

For simplicity's sake, suppose we begin with a culture of one bacterial cell that can divide every hour. Figure 1 shows the number of bacterial cells at the end of each subsequent hour. We see that the single bacterial cell leads to over one thousand bacterial cells in just ten hours! And if we were to extrapolate the table to twenty-four hours, we would have over 16 million cells!

Hour	0	1	2	3	4	5	6	7	8	9	10
Bacteria	1	2	4	8	16	32	64	128	256	512	1024

Figure 1.

In this chapter, we will explore exponential functions, which can be used for modeling growth patterns such as those found in bacteria. Exponential functions have numerous other real-world applications in the fields of finance, archeology, and medicine to name a few. Here are the topics we'll cover in this chapter:

2.1 Properties of Exponents

OVERVIEW

Mathematicians, scientists, and economists commonly encounter very large and very small numbers, but it may not be obvious how common such figures are in everyday life. For example, a digital video camera records many tiny picture elements, called pixels, to form a single image, or frame. A camera might record frames in 2,048 pixels by 1,536 pixels and each pixel might contain 48 bits of information pertaining to the color that pixel should represent. If the camera shoots video at 24 frames per second, the number of bits of information in a one-hour video is extremely large.

We can compute the maximum number of bits of information in our example above. Using a calculator, we enter 2048 × 1536 × 48 × 24 × 3600 and press ENTER. The calculator displays 1.304596316E13. What does this mean? The "E13" portion of the result represents the exponent 13 of ten, so there are a maximum of approximately 1.3×10^{13} bits of data in that one-hour film. In this section, we review rules of exponents first and then apply them to calculations involving very large or small numbers.

As you study this section, you will learn how to:

- Use the five basic properties of exponents.
- Use the zero exponent rule of exponents.
- Use negative integer exponents.
- Simplify exponential expressions.
- Use scientific notation.

A. DEFINITION OF AN EXPONENT

We use exponents as a kind of shorthand for repeated multiplication. If we multiply the number 5 times itself 6 times, we can write $5 \cdot 5 \cdot 5 \cdot 5 \cdot 5 \cdot 5$, but it is much shorter to write 5^6. In this example, we call 5 the **base** and 6 the **exponent**. When written as 5^6, the meaning is that 5 is a factor 6 times. We call 5^6 a **power** of 5, namely the 6th power of 5.

In general, for any natural number n, the notation b^n indicates that b is a factor n times.

For most powers of a base, we use our calculators to find the value. For 5^6, type the base 5 and then use the exponent button $\boxed{\wedge}$ before typing 6. In Figure 1, we see that $5^6 = 15625$.

Notice that b^1 stands for one factor of b, so $b^1 = b$. We typically do not write the exponent in this case. Two other powers have specific names. We refer to b^2 as "b squared" and b^3 as "b cubed." These nicknames come from geometry formulas for the

> ### Exponents
>
> For any natural number n,
>
> $$b^n = b \cdot b \cdot b \cdot \ldots \cdot b$$
>
> where b is a factor n times.
>
> We refer to b^n as "b raised to the nth power." We call b the **base** and n the **exponent**.

area of a square and the volume of a cube.

It's common to confuse the two expressions -3^4 and $(-3)^4$. The first expression indicates that we should raise 3 to the 4th power and then find its opposite.

$$-3^4 = -(3^4) = -(3 \cdot 3 \cdot 3 \cdot 3) = -81$$

While the second expression indicates that we should raise -3 to the 4th power.

$$(-3)^4 = (-3)(-3)(-3)(-3) = 81$$

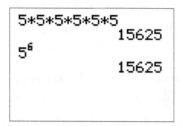

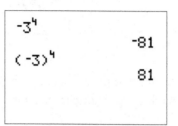

Figure 1. Calculating powers on the calculator.

Figure 2. Compute $-3\ 4$ and $(-3)^4$.

We can use a calculator to check   both computations, as you can see in Figure 2.

B. PROPERTIES OF EXPONENTS

The Product Property of Exponents

Consider the product $x^3 \cdot x^4$. Both terms have the same base, x, but they are raised to different exponents. We expand each expression, and then rewrite the resulting expression.

$$x^3 \cdot x^4 = \underbrace{x \cdot x \cdot x \cdot}_{\text{3 factors}} \underbrace{x \cdot x \cdot x \cdot x}_{\text{4 factors}}$$

$$= \underbrace{x \cdot x \cdot x \cdot x \cdot x \cdot x \cdot x}_{\text{7 factors}}$$

$$= x^7$$

The result is that $x^3 \cdot x^4$ equals x^{3+4}, which equals x^7.

Notice that the exponent of the product is the sum of the exponents of the factors. In other words, when multiplying exponential expressions with the same base, we write the result with the common base and add the exponents. This is the product property of exponents and can be stated in general terms as

$$b^m \cdot b^n = b^{m+n}$$

The Quotient Property of Exponents

In a similar way to the product property, we can simplify an expression such as $\frac{y^m}{y^n}$, where $m > n$. Consider the example $\frac{y^9}{y^5}$. Perform the division by canceling common factors.

$$\frac{y^9}{y^5} = \frac{y \cdot y \cdot y \cdot y \cdot y \cdot y \cdot y \cdot y \cdot y}{y \cdot y \cdot y \cdot y \cdot y} \qquad \text{Definition of exponent.}$$

$$= \frac{\cancel{y} \cdot \cancel{y} \cdot \cancel{y} \cdot \cancel{y} \cdot \cancel{y} \cdot y \cdot y \cdot y \cdot y}{\cancel{y} \cdot \cancel{y} \cdot \cancel{y} \cdot \cancel{y} \cdot \cancel{y}} \qquad \text{Divide: } \frac{y}{y} = 1$$

$$= \frac{y \cdot y \cdot y \cdot y}{1} \qquad \text{Simplify.}$$

$$= y^4$$

Notice that the exponent of the quotient is the difference between the exponents of the divisor and dividend. So the shortcut would look like this:

$$\frac{y^9}{y^5} = y^{9-5} = y^4$$

In other words, when dividing exponential expressions with the same base, we write the result with the common base and subtract the exponents. This is the quotient property of exponents, which can be stated in general terms as

$$\frac{b^m}{b^n} = b^{m-n}, \text{ where } m > n$$

For the time being, we must be aware of the condition $m > n$. Otherwise, the difference $m - n$ could be zero or negative. We'll explore those possibilities soon. Also, instead of qualifying variables as nonzero each time, we will simplify matters and assume from here on that all variables represent nonzero real numbers.

All five basic properties of exponents can be demonstrated in a similar way. They are written in general terms below.

The Five Properties of Exponents

For any real number b and natural numbers m and n,

$$b^m \cdot b^n = b^{m+n} \qquad \text{Product property of exponents.}$$

$$\frac{b^m}{b^n} = b^{m-n}, b \neq 0, m > n \quad \text{Quotient property of exponents.}$$

$$(bc)^n = b^n c^n \qquad \text{Raising a product to a power.}$$

$$\left(\frac{b}{c}\right)^n = \frac{b^n}{c^n}, c \neq 0 \qquad \text{Raising a quotient to a power.}$$

$$(b^m)^n = b^{mn} \qquad \text{Raising a power to a power.}$$

▶ Example 1

Using the Product and Quotient Properties

Simplify the following expressions.

1. $b^5 \cdot b^3$

2. $(-2)^5 \cdot (-2)$

3. $x^2 \cdot x^5 \cdot x^3$

4. $\dfrac{b^{18}}{b^{11}}$

5. $\dfrac{7^{11}}{7^9}$

6. $\dfrac{24\,a^6}{3\,a^5}$

Solutions

1. $b^5 \cdot b^3 = b^{3+5} = b^8$

2. $(-2)^5 \cdot (-2) = (-2)^5(-2)^1 = (-2)^6 = 64$

3. We can extend the product property of exponents to simplify a product of three factors having the same base. Simply add the exponents of all three factors:

$$x^2 \cdot x^5 \cdot x^3 = x^{2+5+3} = x^{10}$$

4. $\dfrac{b^{18}}{b^{11}} = b^{18-11} = b^7$

5. $\dfrac{7^{11}}{7^9} = 7^{11-9} = 7^2 = 49$

6. $\dfrac{24\,a^6}{3\,a^5} = \dfrac{24}{3}\,a^{6-5} = 8a$

We can use all five properties of exponents in a variety of combinations to simplify more complex expressions. At first, it's wise to proceed methodically using one property at a time. With practice, you may find that you're able to combine some of the steps.

▶ Example 2

Simplifying Expressions Involving Exponents

Simplify.

1. $(2b^3 c^6)^4$

2. $(5bc^5)(9\,b^4 c^7)$

3. $\dfrac{4\,b^8 c^9}{20\,b^7 c^5}$

4. $\left(\dfrac{3\,b^5 c^4}{b^3 c\,d^{10}}\right)^3$

Solutions

1.

$$
\begin{aligned}
(2b^3 c^6)^4 &= 2^4 (b^3)^4 (c^6)^4 && \text{Raise factors to the 4th power: } (bc)^n = b^n c^n \\
&= 16\,b^{12} c^{24} && \text{Simplify the numerical part. Multiply exponents: } (b^m)^n = b^{mn}
\end{aligned}
$$

2.

$$
\begin{aligned}
(5bc^5)(9\,b^4 c^7) &= (5 \cdot 9)(b\,b^4)(c^5 c^7) && \text{Rearrange factors.} \\
&= 45\,b^5 c^{12} && \text{Simplify the numerical part. Add exponents: } b^m b^n = b^{m+n}
\end{aligned}
$$

3.

$$\frac{4\,b^8\,c^9}{20\,b^7\,c^5} = \frac{1}{5}b^1\,c^4 \qquad \text{Simplify numeric fraction. Subtract exponents: } \frac{b^m}{b^n} = b^{m-n}$$

$$= \frac{b\,c^4}{5} \qquad \text{Simplify.}$$

4.

$$\left(\frac{3\,b^5\,c^4}{b^3\,c\,d^{10}}\right)^3 = \left(\frac{3\,b^2\,c^3}{d^{10}}\right)^3 \qquad \text{Subtract exponents: } \frac{b^m}{b^n} = b^{m-n}$$

$$= \frac{(3\,b^2\,c^3)^3}{(d^{10})^3} \qquad \text{Raise numerator and denominator to the 3rd power: } \left(\frac{b}{c}\right)^n = \frac{b^n}{c^n}$$

$$= \frac{3^3\,(b^2)^3\,(c^3)^3}{(d^{10})^3} \qquad \text{Raise factors to the 3rd power: } (bc)^n = b^n\,c^n$$

$$= \frac{27\,b^6\,c^9}{d^{30}} \qquad \text{Multiply exponents: } (b^m)^n = b^{mn}$$

Practice Set B

Simplify the following expressions. Turn the page to check your work.

1. $t^3 \cdot t^6 \cdot t^5$

2. $(6\,a^6\,b^3)(4\,a^9\,b^2)$

3. $\dfrac{33\,a^9\,b^5}{3\,a^2\,b^2}$

4. $(4x^2y^7)^5$

5. $\left(\dfrac{2x^3y^{13}}{10xy^{10}}\right)^4$

C. ZERO EXPONENTS AND NEGATIVE EXPONENTS

The Zero Exponent Rule of Exponents

Let's return to the quotient rule. We made the condition that $m > n$ so that the difference $m - n$ would never be zero or negative. What would happen if $m = n$? For example, let's simplify the expression $\frac{b^8}{b^8}$. If we were to simplify this expression using the quotient property, we would have this:

The Zero Exponent Rule
For any nonzero real number b,
$b^0 = 1$

$$\frac{b^8}{b^8} = b^{8-8} = b^0$$

If we view the expression $\frac{b^8}{b^8}$ in a different way, we remember that any number divided by itself is equal to 1. That means:

$$\frac{b^8}{b^8} = 1$$

If we equate the two answers, the result is $b^0 = 1$. This is true for any nonzero real number, or any variable representing such a number.

▶ *Example 3*

Using the Zero Exponent Rule

Simplify each expression.

1. $\dfrac{a^0 b^4 c^5}{c^5}$
2. $(7 a^3 b^0)^2$
3. $9 + a^0 + b^0$

Solution

1. $\dfrac{a^0 b^4 c^5}{c^5} = a^0 b^4 c^0 = (1) b^4 (1) = b^4$

Recall that any number times 1 equals itself ($1a = a$) and that we do not need to write the number 1 when it is a coefficient.

2. $(7 a^3 b^0)^2 = (7 a^3 \cdot 1)^2 = (7 a^3)^2 = 49 a^6$

Notice in the last step we are squaring the 7 as well as the variable term because it is in parentheses.

3. $9 + a^0 + b^0 = 9 + 1 + 1 = 11$

Be careful! Adding 1 to a term is not the same as multiplying by 1.

The Negative Integer Rule of Exponents

Another useful result occurs if we relax the condition that $m > n$ in the quotient property even further and let $m < n$. For example, let's simplify $\dfrac{b^3}{b^5}$. Begin by using the definition of an exponent.

$$\dfrac{b^3}{b^5} = \dfrac{b \cdot b \cdot b}{b \cdot b \cdot b \cdot b \cdot b} \qquad \text{Definition of exponent.}$$

$$\dfrac{b^3}{b^5} = \dfrac{\cancel{b} \cdot \cancel{b} \cdot \cancel{b}}{\cancel{b} \cdot \cancel{b} \cdot \cancel{b} \cdot b \cdot b} \qquad \text{Divide: } \dfrac{b}{b} = 1$$

$$= \dfrac{1}{b \cdot b} \qquad \text{Simplify.}$$

$$= \dfrac{1}{b^2}$$

If we simplify the original expression using the quotient property, we have

$$\dfrac{b^3}{b^5} = b^{3-5} = b^{-2}$$

Putting the answers together, we have $b^{-2} = \dfrac{1}{b^2}$. This is true for any nonzero real number, or any variable representing a nonzero real number.

For the quotient $\dfrac{b^m}{b^n}$, when $m < n$, we can use the negative rule of exponents to simplify the expression to its reciprocal because the difference $m - n$ is negative. A factor with a negative exponent becomes the same factor with a positive exponent if we move it across the fraction bar, from numerator to denominator, or vice versa.

$$b^{-n} = \dfrac{1}{b^n} \text{ and } \dfrac{1}{b^{-n}} = b^n$$

> ### The Negative Integer Rule of Exponents
>
> For any nonzero real number b and natural number n, the negative rule of exponents states that
>
> $$b^{-n} = \dfrac{1}{b^n} \text{ and } \dfrac{1}{b^{-n}} = b^n$$

▶ *Example 4*

Using the Negative Exponent Rule

Simplify. Write answers with positive exponents.

1. $\dfrac{c^3}{c^{10}}$ **2.** $b^2 \cdot b^{-8}$ **3.** $\dfrac{4\,a^{-5}}{b^{-9}}$ **4.** $\dfrac{-3\,a^8\,b^{-7}}{c^{-1}}$

Solutions

1. We use the quotient property in the first step and the negative exponent rule in the last step.

$$\frac{c^3}{c^{10}} = c^{3-10} = c^{-7} = \frac{1}{c^7}$$

2. We use the product property in the first step and the negative exponent rule in the last step.

$$b^2 \cdot b^{-8} = b^{2+8} = b^{-6} = \frac{1}{b^6}$$

3. A factor with a negative exponent becomes the same factor with a positive exponent if it is moved across the fraction bar.

$$\frac{4\,a^{-5}}{b^{-9}} = \frac{4\,b^9}{a^5}$$

4. Be careful! The coefficient, −3, is a negative number, not a negative exponent.

$$\frac{-3\,a^8\,b^{-7}}{c^{-1}} = \frac{-3\,a^8\,c}{b^7}$$

Practice Set C

Simplify. Write answers with positive exponents. Turn the page to check your work.

6. $(a^3\,b\,c^0)(b^9\,c^2)$ **8.** $\dfrac{2\,b^4\,c^3}{b^7\,c^3}$ **10.** $\dfrac{b^{-9}\,c^2}{a^7\,d^{-4}}$

7. $a^0\,b^8 + b^0$ **9.** $\dfrac{-9\,a^{13}\,c^5}{a^9\,c^{12}}$

Practice Set B — Answers

1. $t^3 \cdot t^6 \cdot t^5 = t^{14}$

2. $(6\,a^6\,b^3)(4\,a^9\,b^2) = (6 \cdot 4)\,a^{6+9}\,b^{3+2} = 24\,a^{15}\,b^5$

3. $\dfrac{33\,a^9\,b^5}{3\,a^2\,b^2} = \dfrac{33}{3}\,a^{9-2}\,b^{5-2} = 11\,a^7\,b^3$

4. $(4x^2y^7)^5 = 4^5(x^2)^5(y^7)^5 = 1024\,x^{10}\,y^{35}$

5. $\left(\dfrac{2x^3y^{13}}{10xy^{10}}\right)^4 = \left(\dfrac{1\,x^2\,y^3}{5}\right)^4$

$= \dfrac{(x^2)^4\,(y^3)^4}{5^4} = \dfrac{x^8\,y^{12}}{625}$

D. SIMPLIFYING EXPRESSIONS INVOLVING EXPONENTS

We have seen that the exponential expression b^n is defined when n is a natural number, 0, or the negative of a natural number. That means that b^n is defined for any integer n and therefore all of the exponent properties we have seen in this section hold for any integer n. Furthermore, we can combine any of these properties with the rules for zero and negative exponents to simplify complex expressions.

> ### Simplifying Expressions Involving Exponents
>
> An expression involving exponents is simplified if
>
> ♦ It includes no parentheses.
>
> ♦ Each variable or constant appears as a base as few times as possible.
>
> ♦ Each numerical expression has been calculated and each numerical fraction has been written in lowest terms.
>
> ♦ Each exponent is positive.

▶ *Example 5*

Simplifying Exponential Expressions

Simplify each expression.

1. $(6\,m^2\,n^{-1})^3$

2. $17^5 \cdot 17^{-4} \cdot 17^{-3}$

3. $\left(\dfrac{u^{-1}v}{v^{-1}}\right)^2$

4. $(-2\,a^3\,b^{-1})(5\,a^{-2}\,b^2)$

5. $(x^2\sqrt{2})^4(x^2\sqrt{2})^{-4}$

6. $\dfrac{(3\,w^2)^5}{(6\,w^{-2})^2}$

Solutions

1.

$$
\begin{aligned}
(6\,m^2\,n^{-1})^3 &= (6)^3(m^2)^3(n^{-1})^3 && \text{Product property.}\\
&= 6^3\,m^{2\cdot3}\,n^{-1\cdot3} && \text{Raise a power to a power.}\\
&= 216\,m^6\,n^{-3} && \text{Simplify.}\\
&= \frac{216\,m^6}{n^3} && \text{Apply the negative exponent rule.}
\end{aligned}
$$

2.

$$
\begin{aligned}
17^5 \cdot 17^{-4} \cdot 17^{-3} &= 17^{5-4-3} && \text{Product property.}\\
&= 17^{-2} && \text{Simplify.}\\
&= \frac{1}{17^2} \text{ or } \frac{1}{289} && \text{Apply the negative exponent rule.}
\end{aligned}
$$

3.

$$
\begin{aligned}
\left(\frac{u^{-1}v}{v^{-1}}\right)^2 &= \frac{(u^{-1}v)^2}{(v^{-1})^2} && \text{Raise a quotient to a power.}\\
&= \frac{u^{-2}v^2}{v^{-2}} && \text{Raise a product to a power.}\\
&= u^{-2}\,v^{2-(-2)} && \text{Apply the quotient property.}\\
&= u^{-2}v^4 && \text{Simplify.}\\
&= \frac{v^4}{u^2} && \text{Apply the negative exponent rule.}
\end{aligned}
$$

4.

$$(-2a^3b^{-1})(5a^{-2}b^2) = -2 \cdot 5 \cdot a^3 \cdot a^{-2} \cdot b^{-1} \cdot b^2 \qquad \text{Rearrange terms.}$$
$$= -10 \cdot a^{3-2} \cdot b^{-1+2} \qquad \text{Product property.}$$
$$= -10ab \qquad \text{Simplify.}$$

5.

$$\left(x^2\sqrt{2}\right)^4 \left(x^2\sqrt{[2]}\right)^{-4} = \left(x^2\sqrt{2}\right)^{4-4} \qquad \text{Product property.}$$
$$= \left(x^2\sqrt{2}\right)^0 \qquad \text{Simplify.}$$
$$= 1 \qquad \text{Apply the zero exponent rule.}$$

6.

$$\frac{(3w^2)^5}{(6w^{-2})^2} = \frac{(3)^5 \cdot (w^2)^5}{(6)^2 \cdot (w^{-2})^2} \qquad \text{Raise a product to a power.}$$
$$= \frac{3^5 w^{2 \cdot 5}}{6^2 w^{-2 \cdot 2}} \qquad \text{Raise a power to a power.}$$
$$= \frac{243 w^{10}}{36 w^{-4}} \qquad \text{Simplify.}$$
$$= \frac{27 w^{10-(-4)}}{4} \qquad \text{Apply the quotient property and then reduce the fraction.}$$
$$= \frac{27 w^{14}}{4} \qquad \text{Simplify.}$$

Practice Set D

Simplify each expression. Turn the page to check your work.

11. $(2uv^{-2})-3$

12. $x^8 \cdot x^{-12} \cdot x$

13. $\left(\frac{a^2 b^{-3}}{b^{-1}}\right)^2$

14. $(9r^{-5}s^3)(3r^6s^{-4})$

15. $\left(\frac{4}{9}tw^{-2}\right)^{-3}\left(\frac{4}{9}tw^{-2}\right)^3$

16. $\frac{(2h^2k)^4}{(7h^{-1}k^2)^2}$

Practice Set C — Answers

6. $(a^3bc^0)(b^9c^2) = a^3b^9c^2$

7. $a^0b^8 + b^0 = b^8 + 1$

8. $\frac{2b^4c^3}{b^7c^3} = \frac{2}{b^3}$

9. $\frac{-9a^{13}c^5}{a^9c^{12}} = \frac{-9a^4}{c^7}$

10. $\frac{b^{-9}c^2}{a^7d^{-4}} = \frac{c^2d^4}{a^7b^9}$

E. SCIENTIFIC NOTATION

At the beginning of this section, we used the number 1.3×10^{13} to describe the bits of information in digital images. Other extreme numbers include the width of your instructor's hair, which is about 0.00005 m, and the radius of an electron, which is about 0.00000000000047 m. How can we effectively read, compare, and calculate with numbers like these?

> ### Scientific Notation
>
> A number is written in scientific notation if it is written in the form $a \times 10^{n}$, where $1 \le |a| < 10$ and n is an integer.

A shorthand method of writing very small and very large numbers is called **scientific notation**. This expresses numbers in terms of exponents of 10. To write a number in scientific notation, then, you move the decimal point to the right of the first non-zero digit in the number until you can write the digits as a decimal number between 1 and 10. Count the number of places, n, that you moved the decimal point. You then multiply the decimal number by 10 raised to a power of n. The sign of n (positive or negative) depends on whether you moved the decimal point in the standard form of the number to the right or to the left.

If you move the decimal point left, as with a large number, then n is positive. If you move the decimal point right, as with a small number, then n is negative. For example, consider the number 2,780,418. This is a whole number so the decimal point is understood to be to the right of the digit in the ones column.

$$2,780,418 = 2,780,418.0$$

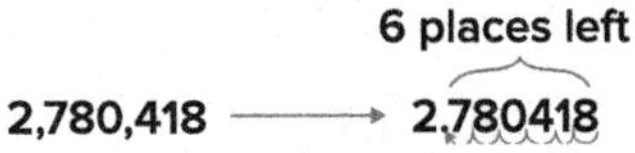

6 places left

2,780,418 $\longrightarrow$ 2.780418

Figure 3.

Move the decimal point to the left until it is to the right of the first nonzero digit, which is 2.

We obtain 2.780418 by moving the decimal point 6 places to the left. Therefore, the exponent of 10 is 6. It's a positive exponent because the value of the original number is greater than 1:

$$2,780,418 = 2.780418 \times 10^{6}$$

Working with small numbers is similar. Take, for example, the radius of an electron, 0.00000000000047 m. Perform the same series of steps as above, except move the decimal point to the right. Be careful to *not* include the leading 0 in your count.

13 places right

0.00000000000047 $\longrightarrow$ 00000000000004.7

Figure 4.

We move the decimal point 13 places to the right, so the exponent of 10 is −13. The exponent is negative because the value of the original number is less than 1:

$$0.00000000000047 = 4.7 \times 10^{-13}$$

▶ *Example 6*

Converting Standard Notation to Scientific Notation

Write each number in scientific notation.

1. Distance to Andromeda Galaxy from Earth: 24,000,000,000,000,000,000,000m

2. Diameter of Andromeda Galaxy: 1,300,000,000,000,000,000,000m

3. Number of stars in Andromeda Galaxy: 1,000,000,000,000

4. Diameter of electron: 0.00000000000094m

5. Probability of being struck by lightning in any single year: 0.00000143

Solutions

1. Move the decimal point to the left 22 places:

$$24,000,000,000,000,000,000,000 = 2.4 \times 10^{22}$$

2. Move the decimal point to the left 21 places:

$$1,300,000,000,000,000,000,000 = 1.3 \times 10^{21}$$

3. Move the decimal point to the left 12 places:

$$1,000,000,000,000 = 1 \times 10^{12}$$

4. Move the decimal point to the right 13 places:

$$0.00000000000094 = 9.4 \times 10^{-13}$$

5. Move the decimal point to the right 6 places:

$$0.00000143 = 1.43 \times 10^{-6}$$

Notice that if the number in standard notation is greater than 1, as you see in examples 1–3, the exponent of 10 is positive. If the number in standard notation is less than 1, as you see in examples 4–5, the exponent of 10 is negative.

Converting from Scientific to Standard Notation

To convert a number in **scientific notation** to standard notation, simply reverse the process. Move the decimal point n places to the right if n is positive or n places to the left if n is negative. Add zeros as needed. Remember, if n is positive, the value of the number is *greater* than 1, and if n is negative, the value of the number is *less* than one.

Practice Set D — Answers

11. $\dfrac{v^6}{8u^3}$	13. $\dfrac{a^4}{b^4}$	15. 1
12. $\dfrac{1}{x^3}$	14. $\dfrac{27r}{s}$	16. $\dfrac{16h^{10}}{49}$

▶ *Example 7*

Converting Scientific Notation to Standard Notation

Convert each number in scientific notation to standard notation.

1. 3.547×10^{14} 3. 7.91×10^{-7}

2. -2×10^{6} 4. -8.05×10^{-12}

Solutions

1. Move the decimal point 14 places to the right:
$$3.547 \times 10^{14} = 354{,}700{,}000{,}000{,}000$$

2. Move the decimal point 6 places to the right:
$$-2 \times 10^{6} = -2{,}000{,}000$$

3. Move the decimal point 7 places to the left:
$$7.91 \times 10^{-7} = 0.000000791$$

4. Move the decimal point 12 places to the left:
$$-8.05 \times 10^{-12} = -0.00000000000805$$

Using Scientific Notation in Applications

When used with the rules of exponents, scientific notation makes calculating with large or small numbers much easier than doing so using standard notation. For example, suppose we are asked to calculate the number of atoms in 1 liter of water. Each water molecule contains 3 atoms (2 hydrogen and 1 oxygen). The average drop of water contains around 1.32×10^{21} molecules of water, and 1 liter of water holds about 1.22×10^{4} average drops. Therefore, there are approximately $3 \bullet (1.32 \times 10^{21}) \bullet (1.22 \times 10^{4})$ $\approx 4.83 \times 10^{25}$ atoms in 1 liter of water. We simply multiply the decimal terms and add the exponents. Imagine having to perform the calculation without using scientific notation!

When performing calculations with scientific notation, be sure to write the answer in proper scientific notation. For example, consider the product $(7 \times 10^{4}) \bullet (5 \times 10^{6}) = 35 \times 10^{10}$. The answer is not in proper scientific notation because 35 is greater than 10. Consider $35 = 3.5 \times 10^{1}$. That adds a one to the exponent of the answer:

$$
\begin{aligned}
(35) \times 10^{10} &= (3.5 \times 10^{1}) \times 10^{10} \\
&= 3.5 \times (10^{1} \times 10^{10}) \\
&= 3.5 \times 10^{11}
\end{aligned}
$$

▶ *Example 8*

Using Scientific Notation

Perform the operations and write the answer in scientific notation.

1. $(8.14 \times 10^{-7})(6.5 \times 10^{10})$ **3.** $(2.7 \times 10^5)(6.04 \times 10^{13})$

2. $(4 \times 10^5) \div (-1.52 \times 10^9)$ **4.** $(1.2 \times 10^8) \div (9.6 \times 10^5)$

Solutions

1.

$$
\begin{aligned}
(8.14 \times 10^{-7})(6.5 \times 10^{10}) &= (8.14 \times 6.5)(10^{-7} \times 10^{10}) && \text{Rearrange factors.} \\
&= (52.91)(10^3) && \text{Product property of exponents.} \\
&= 5.291 \times 10^1 \times 10^3 && \text{Convert to scientific notation.} \\
&= 5.291 \times 10^4 && \text{Product property of exponents.}
\end{aligned}
$$

2.

$$
\begin{aligned}
(4 \times 10^5) \div (-1.52 \times 10^9) &= \left(\frac{4}{-1.52}\right)\left(\frac{10^5}{10^9}\right) && \text{Rearrange factors.} \\
&= (-2.63)(10^{-4}) && \text{Quotient property of exponents.} \\
&= -2.63 \times 10^{-4} && \text{Convert to scientific notation.}
\end{aligned}
$$

3.

$$
\begin{aligned}
(2.7 \times 10^5)(6.04 \times 10^{13}) &= (2.7 \times 6.04)(10^5 \times 10^{13}) && \text{Rearrange factors.} \\
&= (16.308)(10^{18}) && \text{Product property of exponents.} \\
&= 1.6308 \times 10^1 \times 10^{18} && \text{Convert to scientific notation.} \\
&= 1.6308 \times 10^{19} && \text{Product property of exponents.}
\end{aligned}
$$

4.

$$
\begin{aligned}
(1.2 \times 10^8) \div (9.6 \times 10^5) &= \left(\frac{1.2}{9.6}\right)\left(\frac{10^8}{10^5}\right) && \text{Rearrange factors.} \\
&= (0.125)(10^3) && \text{Quotient property of exponents.} \\
&= 1.25 \times 10^{-1} \times 10^3 && \text{Convert to scientific notation.} \\
&= 1.25 \times 10^2 && \text{Product property of exponents.}
\end{aligned}
$$

Practice Set E

Write each number in scientific notation.

17. U.S. national debt per taxpayer (April 2014): $152,000

18. World population (April 2014): 7,158,000,000

19. World gross national income (April 2014): $85,500,000,000,000

20. Time for light to travel 1 m: 0.00000000334 s

21. Probability of winning lottery (match 6 of 49 possible numbers): 0.0000000715

Convert each number in scientific notation to standard notation.

22. 7.03×10^5

23. -8.16×10^{11}

24. -3.9×10^{-13}

25. 8×10^{-6}

Perform the operations and write the answer in scientific notation.

26. $(-7.5 \times 10^8)(1.13 \times 10^{-2})$

27. $(1.24 \times 10^{11}) \div (1.55 \times 10^{18})$

28. $(3.72 \times 10^9)(8 \times 10^3)$

29. $(9.933 \times 10^{23}) \div (-2.31 \times 10^{17})$

When you're finished, turn the page to check your answers.

Exercises

1. Is 2^3 the same as 3^2? Explain.

2. When can you add two exponents?

3. What is the purpose of scientific notation?

4. Explain what a negative exponent does.

For the following exercises, simplify the given expression. Write answers with positive exponents.

5. 9^2

6. 15^{-2}

7. $3^2 \times 3^3$

8. 4^4

9. $(2^2)^{-2}$

10. $(5-8)^0$

11. $11^3 \div 11^4$

12. $6^5 \cdot 6^{-7}$

13. $(8^0)^2$

14. $5^{-2} \div 5^2$

For the following exercises, write each expression with a single base. Do not simplify further. Write answers with positive exponents.

15. $4^2 \cdot 4^3 \div 4^{-4}$

16. $\frac{6^{12}}{6^9}$

17. $(12^3 \cdot 12)^{10}$

18. $10^6 \div (10^{10})^{-2}$

19. $7^{-6} \cdot 7^{-3}$

20. $(3^3 \div 3^4)^5$

For the following exercises, express the decimal in scientific notation.

21. 0.0000314

22. 148,000,000

For the following exercises, convert each number in scientific notation to standard notation.

23. 1.6×10^{10}

24. 9.8×10^{-9}

For the following exercises, simplify the given expression. Write answers with positive exponents.

25. $\frac{a^3 a^2}{a}$

26. $\frac{m n^2}{m^{-2}}$

27. $(b^3 c^4)^2$

28. $\left(\frac{x^{-3}}{y^2}\right)^{-5}$

29. $a b^2 \div d^{-3}$

30. $(w^0 x^5)^{-1}$

31. $\frac{m^4}{n^0}$

32. $y^{-4} (y^2)^2$

33. $\frac{p^{-4} q^2}{p^2 q^{-3}}$

34. $(l \bullet w)^2$

35. $(y^7)^3 \div x^{14}$

36. $\left(\frac{a}{2^3}\right)^2$

37. $5^2 m \div 5^0 m$

38. $\frac{(16 \sqrt{x})^2}{y^{-1}}$

39. $\frac{2^3}{(3a)^{-2}}$

40. $(m a^6)^2 \frac{1}{m^3 a^2}$

41. $(b^{-3} c)^3$

42. $(x^2 y^{13} \div y^0)^2$

43. $(9 z^3)^{-2} y$

44. To reach escape velocity from Earth, the Millennium Falcon must travel at the rate of 2.2×10^6 ft/min. Rewrite the rate in standard notation.

45. A dime is the thinnest coin in U.S. currency. A dime's thickness measures 2.2×10^{-6} m. Rewrite the number in standard notation.

46. The average distance between Earth and the Sun is 92,960,000 mi. Rewrite the distance using scientific notation.

47. A terabyte is made of approximately 1,099,500,000,000 bytes. Rewrite in scientific notation.

48. The Gross Domestic Product (GDP) for the United States in the first quarter of 2014 was 1.71496×10^{13}. Rewrite the GDP in standard notation.

49. One picometer is approximately 3.397×10^{-11} in. Rewrite this length using standard notation.

50. The value of the services sector of the U.S. economy in the first quarter of 2012 was $10,633.6 billion. Rewrite this amount in scientific notation.

51. Avogadro's constant is used to calculate the number of particles in a mole (not the animal). In chemistry, a mole is a unit to measure the amount of a substance. The constant is 6.0221413×10^{23}. Write Avogadro's constant in standard notation.

52. Planck's constant is an important unit of measure in quantum physics describing the relationship between energy and frequency. The constant is written as $6.62606957 \times 10^{-34}$. Write Planck's constant in standard notation.

For the following exercises, simplify the given expression. Write answers with positive exponents.

53. $\left(\frac{3^2}{a^3}\right)^{-2} \left(\frac{a^4}{2^2}\right)^2$

54. $(6^2 - 24)^2 \div \left(\frac{x}{y}\right)^{-5}$

55. $\frac{m^2 n^3}{a^2 c^{-3}} \cdot \frac{a^{-7} n^{-2}}{m^2 c^4}$

56. $\left(\frac{x^6 y^3}{x^3 y^{-3}} \cdot \frac{y^{-7}}{x^{-3}}\right)^{10}$

57. $\left(\frac{(a b^2 c)^{-3}}{b^{-3}}\right)^2$

Practice Set E1— Answers

17. 1.52×10^5	22. 703,000	27. 8×10^{-8}
18. 7.158×10^9	23. –816,000,000,000	28. 2.976×10^{13}
19. 8.55×10^{13}	24. –0.00000000000039	29. -4.3×10^6
20. 3.34×10^{-9}	25. 0.000008	
21. 7.15×10^{-8}	26. -8.475×10^6	

2.2 Rational Exponents

OVERVIEW

In Section 2.1, we worked with exponents that are integers. In this section, we will investigate the meaning of rational exponents — exponents that are written as fractions in lowest terms. As you study this section, you will learn how to:

♦ Interpret rational exponents.

♦ Evaluate expressions with rational exponents.

♦ Simplify expressions that have rational exponents.

A. RATIONAL EXPONENTS WITH UNIT FRACTIONS

To develop a definition of fractional exponents, we begin by examining the expression $b^{1/2}$. This expression should obey the same rules for exponents that we studied in Section 2.1, so we use the product property to simplify the following expression:

$$b^{1/2} \cdot b^{1/2} = b^{1/2+1/2} = b^1 = b$$

This suggests that $b^{1/2}$ is a number that when multiplied times itself ($b^{1/2} \cdot b^{1/2}$) produces b. A number multiplied by itself that yields b as a product is known as the square root of b. Thus it is reasonable to define $b^{1/2}$ as $\sqrt{b}$, the **principal square root** of b. So if $b = 9$, then:

$$9^{1/2} \cdot 9^{1/2} = 9^{1/2+1/2} = 9^1 = 9$$

That implies that $9^{1/2} = \sqrt{9} = 3$.

We can generalize the meaning of fractional exponents in order to define $b^{1/n}$ for any natural number n. This time we use the power property of exponents to simplify the expression:

> ### Definition of $b^{1/n}$
>
> For the natural number n, where $n \neq 1$,
>
> ♦ If n is odd, then $b^{1/n}$ is the number whose nth power is b, called the nth root of b.
>
> ♦ If n is even and $b \geq 0$, then $b^{1/n}$ is the nonnegative number whose nth power is b, called the principal nth root of b.
>
> ♦ If n is even and $b < 0$, then $b^{1/n}$ is not a real number
>
> ♦ $b^{1/n}$ is equal to the expression $\sqrt[n]{b}$.

$$(b^{1/n})^n = b^{n/n} = b^1 = b$$

We use $b^{1/n}$ as a factor n times in the expression $(b^{1/n})^n$. It is thus reasonable to define $b^{1/n}$ as $\sqrt[n]{b}$, the **principal nth root** of b. So if $b = 8$, then

$$(8^{1/3})^3 = 8^{3/3} = 8^1 = 8$$

That implies that $8^{1/3} = \sqrt[3]{8} = 2$. The number 2 is called the third root, or **cube root** of 8. If $b = -8$, then $(-8)^{1/3} = -2$ because $(-2)^3 = -8$. However, we cannot assign a real-number value to $(-9)^{1/2}$ because no real number squared is equal to -9.

▶ *Example 1*

Representing and Evaluating *n*th Roots

Rewrite the exponential expressions with radical expressions and evaluate the *n*th root.

1. $49^{1/2}$

2. $27^{1/3}$

3. $(-125)^{1/3}$

4. $10000^{1/4}$

5. $32^{1/5}$

6. $(-16)^{1/2}$

Solution

1. $49^{1/2} = \sqrt{49} = 7$ because $7^2 = 49$.

2. $27^{1/3} = \sqrt[3]{27} = 3$ because $3^3 = 27$.

3. $(-125)^{1/3} = \sqrt[3]{-125} = -5$ because $(-5)^3 = -125$.

4. $10000^{1/4} = \sqrt[4]{10000} = 10$ because $(10)^4 = 10000$.

5. $32^{1/5} = \sqrt[5]{32} = 2$ because $2^5 = 32$.

6. $(-16)^{1/2} = \sqrt{-16}$, which is not a real number, because the square of any real number is nonnegative.

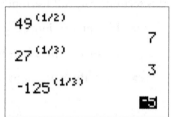

Figure 1. Checks for Problems 1, 2, and 3.

We can check our solutions using a graphing calculator. See Figure 1 for solutions to the first three problems in Example 1. If the expression entered into the calculator produces a non-real number, such as in problem 6 of the example, the calculator will give you an error message. See Figure 2.

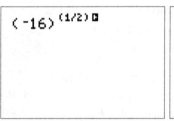

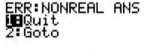

Figure 2. Error message for non-real numbers.

Practice Set A

Rewrite the exponential expressions with radical expressions and evaluate the *n*th root. Check your solutions on your calculator and then turn the page to confirm them.

1. $64^{1/2}$

2. $64^{1/3}$

3. $64^{1/6}$

4. $-64^{1/2}$

5. $(-64)^{1/2}$

6. $(-64)^{1/3}$

B. DEFINITION OF RATIONAL EXPONENTS

We can also have fractional exponents with numerators other than 1. In these cases, the exponent must be a fraction in lowest terms. We then raise the base to a power and take an *n*th root. The numerator tells us the power and the denominator tells us the root.

$$b^{m/n} = \sqrt[n]{b^m} = \left(\sqrt[n]{b}\right)^m$$

All of the properties of exponents that we learned for integer exponents also hold for rational exponents. We express these **rational exponents** in terms of the unit fractions we have already seen in this section.

Notice that we can break up a rational exponent in two different ways. When simplifying, it's usually easier to take the *n*th root of the base first and then raise that root to the power. For example, $9^{3/2}$ can be written as $(9^3)^{1/2}$ or $(9^{1/2})^3$. Below we simplify each expression:

$$(9^3)^{1/2} = 729^{1/2} = \sqrt{729} = 27$$
$$(9^{1/2})^3 = \left(\sqrt{9}\right)^3 = 3^3 = 27$$

The calculation is easier to do — especially *without* a calculator — when we use the second form of the expression. We can check our work on a calculator if desired. We just have to make sure we use parentheses where needed!

> ### Rational Exponents
>
> For the natural numbers *m* and *n*, where $n \neq 1$ and $b^{1/n}$ is a real number,
> - $b^{m/n} = (b^{1/n})^m = (b^m)^{1/n}$
> - $b^{-m/n} = \frac{1}{b^{m/n}}, b \neq 0$
>
> A power of the form $b^{m/n}$ or $b^{-m/n}$ is said to have a **rational exponent**.

▶ **Example 2**

Evaluating Expressions with Rational Exponents

Evaluate the following expressions, and then check your work using a calculator.

1. $16^{3/4}$
2. $25^{3/2}$
3. $(-8)^{5/3}$
4. $(-27)^{2/3}$
5. $36^{-1/2}$
6. $81^{-3/4}$

Solution

1. $16^{3/4} = (16^{1/4})^3 = \left(\sqrt[4]{16}\right)^3 = 2^3 = 8$

2. $25^{3/2} = (25^{1/2})^3 = \left(\sqrt{25}\right)^3 = 5^3 = 125$

3. $(-8)^{5/3} = ((-8)^{1/3})^5 = \left(\sqrt[3]{-8}\right)^5$
   $= (-2)^5 = -32$

4. $(-27)^{2/3} = ((-27)^{1/3})^2 = \left(\sqrt[3]{-27}\right)^2$
   $= (-3)^2 = 9$

5. $36^{-1/2} = \frac{1}{36^{1/2}} = \frac{1}{\sqrt{36}} = \frac{1}{6}$

6. $81^{-3/4} = \frac{1}{81^{3/4}} = \frac{1}{(81^{1/4})^3} = \frac{1}{\left(\sqrt[4]{81}\right)^3}$
   $= \frac{1}{3^3} = \frac{1}{27}$

Don't forget to check your solutions on your calculator!

Practice Set B

Evaluate the following expressions, and then turn the page to check your work.

7. $36^{3/2}$
8. $(-32)^{2/5}$
9. $49^{-1/2}$
10. $9^{-3/2}$
11. $216^{2/3}$
12. $81^{-1/4}$

C. PROPERTIES OF RATIONAL EXPONENTS

All of the properties of exponents that we learned for integer exponents also hold for rational exponents. Next we will apply those rules to simplify expressions involving rational exponents.

▶ *Example 3*

Simplifying Rational Exponents Using the Product Property

Simplify the following expressions. Assume a and b are positive.

   1. $9^{3/8} \cdot 9^{1/8}$ **2.** $(5a^{1/2}b^{4/3})(7a^{1/2}b^{2/3})$ **3.** $3 \cdot b^{2/3} \cdot b^{1/4}$

Solution

1.

$$9^{3/8} \cdot 9^{1/8} = 9^{3/8 + 1/8}$$ Use the product property to add the exponents.

$$= 9^{4/8} = 9^{1/2}$$ Add the fractions to simplify the exponent.

$$= 3$$ 3 is the principal square root of 9.

2.

$$(5a^{1/2}b^{4/3})(7a^{1/2}b^{2/3}) = 35(a^{1/2}b^{4/3})(a^{1/2}b^{2/3})$$ Multiply the coefficients.

$$= 35a^{1/2 + 1/2}b^{4/3 + 2/3}$$ Use the product property to add the exponents.

$$= 35a^{2/2}b^{6/3} = 35ab^2$$ Add the fractions to simplify the exponent.

3.

$$b^{2/3} \cdot b^{1/4} = b^{2/3 + 1/4}$$ Use the product property to add the exponents.

$$= b^{8/12 + 3/12}$$ Use common denominators to add the fractions.

$$= b^{11/12}$$ Add the fractions to simplify the exponent. This expression cannot be simplified further.

Practice Set A — Answers

1. $64^{1/2} = \sqrt{64} = 8$

2. $64^{1/3} = \sqrt[3]{64} = 4$

3. $64^{1/6} = \sqrt[6]{64} = 2$

4. $-64^{1/2} = -\sqrt{64} = -8$
Note that the negative sign is not in parentheses and is applied after the exponent.

5. $(-64)^{1/2} = \sqrt{-64}$, which is not a real number.

6. $(-64)^{1/3} = \sqrt[3]{-64} = -4$

Practice Set B — Answers

7. $36^{3/2} = 6^3 = 216$

8. $(-32)^{2/5} = (-2)^2 = 4$ Notice the use of parentheses until the last step.

9. $49^{-1/2} = \frac{1}{49^{1/2}} = \frac{1}{7}$

10. $9^{-3/2} = \frac{1}{9^{3/2}} = \frac{1}{3^3} = \frac{1}{27}$

11. $216^{2/3} = 6^2 = 36$

12. $81^{-1/4} = \frac{1}{81^{1/4}} = \frac{1}{3}$

▶ *Example 4*

Simplifying Expression Involving Rational Exponents

Simplify the following expressions. Assume a and b are positive.

1. $(8^{5/3})^{2/5}$

2. $\dfrac{15\,b^{5/4}}{3\,b^{3/4}}$

3. $\left(\dfrac{a^{2/5}}{b^4}\right)^{1/2}$

Solution

1.

$$(8^{5/3})^{2/5} = 8^{10/15}$$

Multiply the exponents (power to a power). $\dfrac{5}{3}\cdot\dfrac{2}{5} = \dfrac{10}{15}$

$$= 8^{2/3}$$

Simplify the fraction.

$$= (8^{1/3})^2$$

Definition of rational exponent.

$$= 2^2 = 4$$

Simplify.

2.

$$\dfrac{15\,b^{5/4}}{3\,b^{3/4}} = \dfrac{5\,b^{5/4}}{b^{3/4}}$$

Divide the coefficients.

$$= 5\,b^{5/4 - 3/4}$$

Use the quotient property to subtract the exponents.

$$= 5\,b^{2/4} = 5\,b^{1/2}$$

Simplify the exponent.

3.

$$\left(\dfrac{a^{2/5}}{b^4}\right)^{1/2} = \dfrac{a^{2/5\,\cdot\,1/2}}{b^{4\,\cdot\,1/2}}$$

Raise a quotient to a power by multiplying the exponents.

$$= \dfrac{a^{2/10}}{b^{4/2}}$$

Multiply the fractions.

$$= \dfrac{a^{1/5}}{b^2}$$

Simplify the fractions.

Practice Set C

Use the properties of exponents to simplify the following expressions. Assume a and b are positive. Turn the page to check your work.

13. $(8a)^{1/3}(14\,a^{4/3})$

14. $\dfrac{2\,b^{3/4}}{16\,b^{1/2}}$

15. $(81\,a^2 b)^{-1/2}$

Exercises

Simplify without using a calculator. Then use a graphing calculator to check your work.

1. $4^{1/2}$

2. $125^{1/3}$

3. $81^{1/4}$

4. $49^{3/2}$

5. $8^{4/3}$

6. $16^{3/4}$

7. $32^{2/5}$

8. $27^{4/3}$

9. $27^{-1/3}$

10. $64^{-1/6}$

11. $4^{-3/2}$

12. $36^{-3/2}$

13. $-8^{2/3}$

14. $-32^{3/5}$

15. $(-64)^{1/3}$

16. $(-27)^{2/3}$

17. $3^{2/5} \cdot 3^{4/5}$

18. $2^{1/4} \cdot 2^{5/4}$

19. $(5^{1/2} \cdot 4^{3/2})^2$

20. $(2^{2/3} \cdot 7^{1/3})^3$

21. $\dfrac{2^{7/4}}{2^{3/4}}$

22. $\dfrac{4^{5/2}}{4^{1/2}}$

23. $\dfrac{8^{1/3}}{8^{2/3}}$

24. $\dfrac{27^{2/3}}{27^{5/3}}$

Use the properties of exponents to simplify the following expressions. Assume all variables are positive.

25. $a^{1/5} a^{2/5}$

26. $k^{5/6} k^{7/6}$

27. $b^{1/4} b^{-3/4}$

28. $a^{4/7} a^{-5/7}$

29. $(25 a^6)^{1/2}$

30. $(27 c^{12})^{1/3}$

31. $(10 a^{1/2} b^{3/5})(3 a^{1/2} b^{7/5})$

32. $(7 b^{1/4} c^{2/3})(4 b^{7/4} c^{5/3})$

33. $\dfrac{a^{3/4}}{a^{1/4}}$

34. $\dfrac{2 c^{5/8}}{c^{3/8}}$

35. $\dfrac{14 b^{6/11}}{2 b^{7/11}}$

36. $\dfrac{21 a^{4/7}}{7 a^{5/7}}$

37. $\dfrac{a^{-1/6}}{a^{5/6}}$

38. $\dfrac{b^{-2/3}}{b^{4/3}}$

39. $\dfrac{12 b^{2/5} c^{-3/10}}{18 b^{-4/5} c^{5/10}}$

40. $\dfrac{15 a^{-1/4} b^{3/5}}{20 a^{7/4} b^{-3/5}}$

Practice Set C — Answers

13. $(2 a^{1/3})(14 a^{4/3}) = 28 a^{5/3}$

14. $\dfrac{1}{8} b^{3/4 - 1/2} = \dfrac{b^{1/4}}{8}$

15. $\dfrac{1}{(81 a^2 b)^{1/2}} = \dfrac{1}{9 a b^{1/2}}$

2.3 Exponential Functions

OVERVIEW

India is the second most populous country in the world with a population of about 1.25 billion people in 2013. The population is growing at a rate of about 1.2% each year. If this rate continues, the population of India will exceed China's population by 2031. When populations grow rapidly, we often say that the growth is "exponential." To a mathematician, however, the term "exponential growth" has a very specific meaning. In this section, we will take a look at exponential functions, which model this kind of rapid growth. Along the way, you will learn to:

- Identify exponential functions
- Evaluate exponential functions.
- Sketch the graph of an exponential function.
- Know the graphical significance of a and b for a function of form $f(x) = ab^x$
- Know the base multiplier property.
- Distinguish between exponential growth and decay.

A. DEFINITION OF AN EXPONENTIAL FUNCTION

When we explored linear growth, we observed a constant rate of change: a constant number by which the output increased or decreased for each unit increase in input. For example, in the equation $f(x) = 3x + 4$, the slope tells us that the output increases by 3 each time the input increases by 1. The scenario in the India population example is different because we have a *percent change* per unit time — rather than a constant change — in the number of people. Exponential functions are the perfect tool to model this type of change.

Examples of exponential functions include:

$$f(x) = 2(3)^x$$

$$g(t) = 22(1.045)^t$$

$$h(x) = 16\left(\frac{3}{5}\right)^x$$

Notice that one defining feature of these equations is that the independent variable is in the exponent position. The base is a constant. Contrast this with the function $q(x) = x^2$, where the base is the variable, x, and the exponent, 2, is a constant. Function q is *not* an exponential function.

> ### Exponential Functions
>
> For any real number x, an **exponential function** is a function with the form $f(x) = ab^x$ where
>
> - a is a non-zero real number called the initial value and
> - b is any positive real number such that $b \neq 1$.
>
> The constant b is called the base.

In general, raising a negative base to a rational exponent with an even denominator results in a non-real number. We limit the base b to positive values to ensure that the outputs will be real numbers. For example, $(-9)^{1/2} = \sqrt{-9}$, which is not a real number. We also limit the base b to positive values other than 1. The equation $f(x) = a(1)^x$ would be a constant function because $1^x = 1$ for all values of x.

Evaluating Exponential Functions

To evaluate an exponential function with the form $f(x) = ab^x$, we simply replace x with the given value and calculate according to the correct order of operations. Be aware that it is a common mistake to multiply the a and b values together before raising to the power. Don't do this! The correct order of operations is to apply the exponent before multiplying.

▶ **Example 1**

Evaluating Exponential Functions

For $f(x) = 2(3)^x$ and $g(x) = 10(.25)^x$, find the following.

1. $f(4)$ 3. $g(2)$
2. $f(-2)$ 4. $g(-1)$

Solution

1.   **2.**

$f(x) = 2(3)^x$ $f(-2) = 2(3)^{-2}$ Substitute $x = -2$.
$f(4) = 2(3)^4$ Substitute $x = 4$. $= 2\left(\frac{1}{9}\right)$ Evaluate the power. Recall
$\quad = 2(81)$ Evaluate the power.   that $3^{-2} = \left(\frac{1}{3^2}\right)$
$\quad = 162$ Multiply. $= \frac{2}{9}$ Multiply.

We can check our answers to 1 and 2 using our calculator. To evaluate $g(x)$ for the given values, we will use a calculator.

3. $g(2) = 10(.25)^2 = 0.625$ **4.** $g(-1) = 10(.25)^{-1} = 40$

Practice Set A

1. Let $h(x) = 6(2)^x$. Evaluate without using a calculator, $h(3)$ and $h(-3)$.

2. Let $f(x) = 200(1.08)^x$. Evaluate $f(5)$ and $f(10)$ using a calculator. Round to three decimal places.

When you're finished, turn the page to check your work.

B. GRAPHING EXPONENTIAL FUNCTIONS

When we graph a specific type of function — in this case an exponential function — for the first time, it's useful to create a table of input-output pairs. We'll let x equal small integer values.

▶ *Example 2*

Graphing an Exponential Function with $b > 1$

Graph $f(x) = 2^x$ by hand. Make observations about the behavior of the graph.

Solution

First we make a table of values

x	-3	-2	-1	0	1	2	3
$f(x) = 2^x$	$\frac{1}{8}$	$\frac{1}{4}$	$\frac{1}{2}$	1	2	4	8

Figure 1.

Notice from the table that the output values are positive for all values of x. As x increases by 1 unit, the output value y is multiplied by 2, the base. As x decreases, the output values grow smaller, approaching zero.

We graph f by plotting the ordered pairs from the table in Figure 1. We can connect these points with a smooth curve because the variable x can be any real number – not just integers. See Figure 2.

Now observe the behavior of the graph of the exponential function $f(x) = 2^x$ and highlight some of its key characteristics:

1. The domain is $(-\infty, \infty)$, because the exponent x can be any real number.

2. The graph of f will never touch the x-axis ($y = 0$) because base 2 raised to any exponent is never equal to zero.

3. The range is $(0, \infty)$.

4. The y-intercept is $(0, 1)$.

5. f is an increasing function.

▪ Notice that the graph of the exponential function in Example 2 gets close to the x-axis but never touches it. When the graph of a function gets closer and closer to a horizontal line as the input value either increases or decreases, we call that line a **horizontal asymptote**. For the exponential function $f(x) = 2^x$, the x-axis, described by the equation $y = 0$ is the horizontal asymptote. See Figure 3.

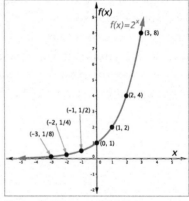

Figure 2. The graph of $f(x) = 2^x$.

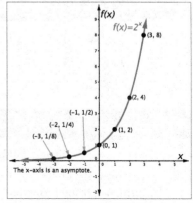

Figure 3. The graph of $f(x) = 2^x$.

▶ *Example 3*

Graphing an Exponential Function with 0 < b < 1

Graph $g(x) = \left(\frac{1}{2}\right)^x$ by hand. Make observations about the behavior of the graph.

Solution

First, in Figure 4, we make a table of values.

x	−3	−2	−1	0	1	2	3
$g(x) = \left(\frac{1}{2}\right)^x$	8	4	2	1	$\frac{1}{2}$	$\frac{1}{4}$	$\frac{1}{8}$

Figure 4.

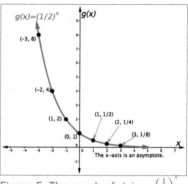

Notice from the table that the output values are positive for all values of x. As x increases, the output values grow smaller, approaching zero. As x decreases, the output values grow larger.

Figure 5 shows the graph of the function, $g(x) = \left(\frac{1}{2}\right)^x$. The domain of $g(x) = \left(\frac{1}{2}\right)^x$ is all real numbers, the range is (0, ∞), and the horizontal asymptote is $y = 0$. The y-intercept is (0, 1).

Figure 5. The graph of $g(x) = \left(\frac{1}{2}\right)^x$.

Although we have only investigated the graphs of two exponential functions, we can generalize our observations to this family of functions as a whole. See the expanded definition of an exponential function to the right.

Expanded Definition of Exponential Function

For any real number x, an exponential function is a function with the form $f(x) = ab^x$ where

◆ a is a non-zero real number called the initial value.

◆ b is any positive real number such that $b \neq 1$.

◆ the domain of f is all real numbers.

◆ the range of f is all positive real numbers if $a > 0$.

◆ the horizontal asymptote is $y = 0$.

Practice Set A — Answers

1. $h(3) = 6(2)^3 = 6(8) = 48$
   $h(-3) = 6(2)^{-3} = 6\left(\frac{1}{2^3}\right)$
   $= 6\left(\frac{1}{8}\right) = \frac{6}{8} = \frac{3}{4}$

2. $f(5) = 200(1.08)^5 \approx 293.866$
   $f(10) = 200(1.08)^{10} \approx 431.785$

Practice Set B

Now it's your turn. When you're finished, turn the page to check your work.

3. Sketch the graph of $f(x) = 4^x$. State the domain, range, and asymptote.

4. Sketch a graph of $f(x) = 0.25^x$. State the domain, range, and asymptote.

C. BASE MULTIPLIER PROPERTY; GROWTH VS. DECAY

The table of values that we used in Example 2 shows that as x increases by 1 unit, y values are multiplied by the base $b = 2$. The table of values that we used in Example 3 shows that as x increases by 1 unit, y values are multiplied by the base $b = \frac{1}{2}$. These observations suggest the **base multiplier property**.

Here are two more examples of the base multiplier property:

1. For the function $y = 2(5)^x$, $b = 5$ so if the value of x increases by 1, the value of y is multiplied by 5.

2. For the function $y = 72(0.8)^x$, $b = 0.8$ so if the value of x increases by 1, the value of y is multiplied by 0.8.

> ### Base Multiplier Property
>
> For an exponential function of the form $y = ab^x$, if the value of the independent variable increases by 1, the value of the dependent variable is multiplied by the base.

When the base b of an exponential function is larger than 1, the function will be increasing. We call this **exponential growth**. When the value of the base is between 0 and 1, the function is decreasing. We call this **exponential decay**. Figure 6 compares the graphs of exponential growth and decay functions.

> ### Exponential Growth and Decay
>
> Let $f(x) = ab^x$, where $a > 0$, $b > 0$, $b \neq 1$. Then
>
> - If $b > 1$, the function is increasing and grows exponentially.
> - If $0 < b < 1$, the function is decreasing and decays exponentially.

When graphing many functions, it is useful to first determine the y-intercept. Let's examine the y-intercepts of functions with the form $f(x) = ab^x$. When a curve crosses the y-axis, the x-value is zero. So in our general form, if $x = 0$, then

> ### y-intercept of an Exponential Function
>
> For an exponential function of the form $y = ab^x$, the y-intercept is $(0, a)$.

$$f(0) = ab^0 = a(1) = a$$

This is because of the zero exponent rule. *Any* base (except 0) to the zero power is equal to 1. This implies that the y-intercept of these functions will be $(0, a)$. For the function $y = 2(5)^x$, the y-intercept is $(0, 2)$. For $y = 72(0.8)^x$, the y-intercept is $(0, 72)$. This seems very simple, but be careful! For the function $y = 3^x$, the y-intercept is *not* $(0, 3)$. If we evaluate the function for $x = 0$, we see $y = 3^0 = 1$, so the y-intercept is $(0, 1)$.

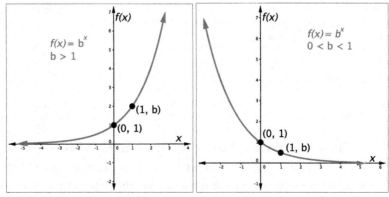

Figure 6. Exponential growth and exponential decay functions..

▶ *Example 4*

For the following exponential functions, determine whether they represent exponential growth or decay. State the *y*-intercept and sketch a rough graph of the function. Check the graph on your graphing calculator.

1. $f(x) = 8(0.65)^x$ **3.** $y = \frac{2}{3}(3)^x$

2. $g(x) = 5(1.5)^x$ **4.** $f(x) = 6\left(\frac{4}{5}\right)^x$

Solution

1. $b = 0.65 < 1$ indicates exponential decay. $a = 8$ means the *y*-intercept is (0, 8).

3. $b = 3 > 1$ indicates exponential growth. $a = \frac{2}{3}$ means the *y*-intercept is (0, ⅔).

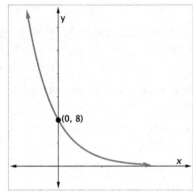

Figure 7.

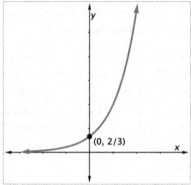

Figure 9.

2. $b = 1.5 > 1$ indicates exponential growth. $a = 5$ means the *y*-intercept is (0, 5).

4. $b = \frac{4}{5} < 1$ indicates exponential decay. $a = 6$ means the *y*-intercept is (0, 6).

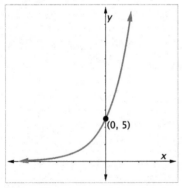

Figure 8.

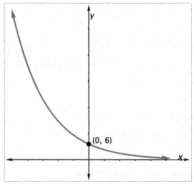

Figure 10.

The exponential functions we will study in this course will all have positive values for *a*. Therefore, whether the function is exponential growth or decay, there will never be an *x*-intercept because there will never be an exponent that makes the base to the power equal to zero. In other words, the equation $b^x = 0$ has no real solutions. However, the exponential functions we study will all have a horizontal asymptote at $y = 0$.

Practice Set C

For the following exponential functions, determine whether they represent exponential growth or decay. State the *y*-intercept and sketch a rough graph of the function. Check the graph on your graphing calculator.

5. $f(x) = 4(1.2)^x$

6. $g(x) = 12(0.75)^x$

When you're finished, turn the page to check your answers.

D, APPLICATIONS OF EXPONENTIAL FUNCTIONS

To differentiate between linear and exponential functions, let's consider two companies, Pancake Warehouse and Carpet Diem. Pancake Warehouse has 100 stores and expands by opening 50 new stores a year, so its growth can be represented by the function $A(x) = 50x + 100$. Carpet Diem has 100 stores and expands by increasing the number of stores by 50% each year, so its growth can be represented by the function $B(x) = 100(1 + 0.5)^x$.

Figure 11 illustrates a few years of growth for these companies.

Year, x	Stores, Pancake Warehouse	Stores, Carpet Diem
0	50(0) + 100 = 100	$100(1 + 0.5)^0 = 100$
1	50(1) + 100 = 150	$100(1 + 0.5)^1 = 150$
2	50(2) + 100 = 200	$100(1 + 0.5)^2 = 225$
3	50(3) + 100 = 250	$100(1 + 0.5)^3 = 337.5$
4	50(4) + 100 = 300	$100(1 + 0.5)^4 = 506.25$

Figure 11.

Practice Set B — Answers

3. The domain is $(-\infty, \infty)$; the range is $(0, \infty)$; the horizontal asymptote is $y = 0$.

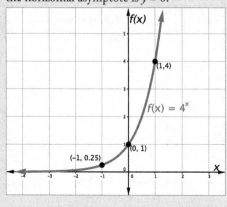

4. The domain is $(-\infty, \infty)$; the range is $(0, \infty)$; the horizontal asymptote is $y = 0$.

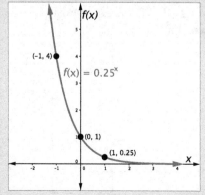

Figure 12 presents a graph comparing the number of stores for each company over a five-year period. We can see that with exponential growth, the number of stores increases much more rapidly than with linear growth.

Notice that the domain for both functions is $[0, \infty)$, and the range for both functions is $[100, \infty)$. The y-intercept for both functions is $(0, 100)$, but after year 1, Carpet Diem always has more stores than Pancake Warehouse.

Let's look more closely at the function representing the number of stores for Carpet Diem, $B(x) = 100(1 + 0.5)^x$. In this exponential function, 100 represents the initial number of stores and $(1 + 0.50)$ represents the original number of stores plus a 50% growth rate. Simplifying the expression in parentheses, we can write this function as $B(x) = 100(1.5)^x$, where 100 is the initial value, and 1.5 is the base.

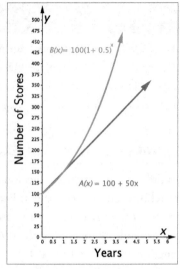

Figure 12. The numbers of stores that Pancake Warehouse and Carpet Diem opened over a five-year period.

▶ Example 5

Evaluating a Real-World Exponential Model

At the beginning of this section, we learned that the population of India was about 1.25 billion in the year 2013, with an annual growth rate of about 1.2%. This situation can be modeled by the exponential growth function $P(t) = 1.25(1.012)^t$, where t is the number of years since 2013. To the nearest thousandth, what will the population of India be in 2031?

Solution

To estimate the population in 2031, we evaluate the model for $t = 18$, because 2031 is 18 years after 2013. Rounding to the nearest thousandth,

$$P(18) = 1.25(1.012)^{18} \approx 1.549$$

There will be about 1.549 billion people in India in the year 2031.

Practice Set C — Answers

5. $b = 1.2 > 1$ indicates exponential growth. $a = 4$ means the y-intercept is $(0, 4)$.

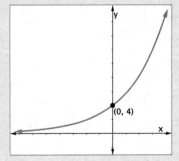

6. $b = 0.75 < 1$ indicates exponential decay. $a = 12$ means the y-intercept is $(0, 12)$.

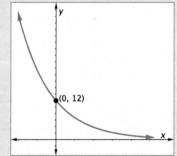

Practice Set D

Try this problem. When you're finished, turn the page to check your work.

7. The population of China was about 1.39 billion in the year 2013, with an annual growth rate of about 0.6%. This situation can be modeled by the function $P(t) = 1.39(1.006)^t$, where t is the number of years since 2013. To the nearest thousandth, predict the population of China for the year 2031. How does this compare to the population prediction we made for India in example 5?

Exercises

1. Explain the base multiplier property.

2. Given a formula for an exponential function, is it possible to determine whether the function grows or decays exponentially just by looking at the formula? Explain.

3. How do you find the y-intercept of an exponential function?

For the following exercises, identify whether the equation represents an exponential function. Explain.

4. $P = f(x) = 13x + 25$

5. $P = f(x) = 13(0.875)^x$

6. $g(t) = 275(1.035)^t$

7. $P = f(x) = 140 - 5x$

8. $h(t) = -4.9t^2 + 18t + 40$

For the following exercises, consider this scenario: For each year t, the population of a forest of trees is represented by the function $A(t) = 115(1.025)^t$. In a neighboring forest, the population of the same type of tree is represented by the function $B(t) = 82(1.029)^t$. (Round answers to the nearest whole number.)

9. Which forest's population is growing at a faster rate?

10. Which forest had a greater number of trees initially? By how many?

11. Assuming the population growth models continue to represent the growth of the forests, which forest will have a greater number of trees after 20 years? By how many?

12. Assuming the population growth models continue to represent the growth of the forests, which forest will have a greater number of trees after 100 years? By how many?

13. Discuss the results from the previous four exercises. Assuming the population growth models continue to represent the growth of the forests, which forest will have the greater number of trees in the long run? Why? What are some factors that might influence the long-term validity of the exponential growth model?

For the following exercises, determine whether the equation represents exponential growth, exponential decay, or neither. Explain.

14. $y = 300(1 - t)^5$

15. $y = 220(1.06)^x$

16. $y = 16.5(1.025)^x$

17. $y = 11,701(0.97)^t$

For the following exercises, graph the function and give the *y*-intercept.

18. $f(x) = 3\left(\frac{1}{2}\right)^x$ **20.** $h(x) = 6(1.75)^x$ **22.** $g(x) = 4(1.5)^x$

19. $g(x) = 2(0.25)^x$ **21.** $f(x) = 5(3)^x$

Use the functions $f(x) = 10(1.2)^x$ and $g(x) = 48(0.95)^x$ for problems 23 – 30. Round results to 3 decimal places.

23. Find $f(4)$.   **26.** Find $f(-3)$ **29.** Find $g(-1)$

24. Find $f(5)$ **27.** Find $g(10)$ **30.** Find $g(-2.5)$

25. Find $f(-2)$ **28.** Find $g(7)$

For the following exercises, graph each set of functions on the same axes.

31. $f(x) = 3\left(\frac{1}{4}\right)^x$, $g(x) = 3(2)^x$, $h(x) = 3(4)^x$

32. $f(x) = \frac{1}{4}(3)^x$, $g(x) = 2(3)^x$, and $h(x) = 4(3)^x$

For the following exercises, match each function with one of the graphs in Figure 13.

33. $f(x) = 2(0.69)^x$ **36.** $f(x) = 4(1.28)^x$

34. $f(x) = 2(1.28)^x$ **37.** $f(x) = 2(1.59)^x$

35. $f(x) = 2(0.81)^x$ **38.** $f(x) = 4(0.69)^x$

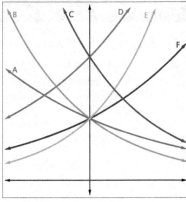

Figure 13.

For the following exercises, use the graphs shown in Figure 14. All have the form $f(x) = ab^x$

39. Which graph has the largest value for *b*?

40. Which graph has the smallest value for *b*?

41. Which graph has the largest value for *a*?

42. Which graph has the smallest value for *a*?

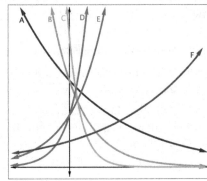

Figure 14.

Practice Set D — Answers

7. To estimate the population in 2031, we evaluate the model for $t = 18$, because 2031 is 18 years after 2013. Rounding to the nearest thousandth,

$$P(18) = 1.39(1.006)^{18} \approx 1.548$$

The model predicts about 1.548 billion people in China in the year 2031. In that year, India's population will exceed China's by about 0.001 billion or 1 million people.

2.4 Finding Equations of Exponential Functions

OVERVIEW

We continue our study of exponential functions by learning to write equations for these functions. Just as you saw with linear models, we have different methods for writing equations. The method to use depends on the information given in the problem. In this section, you will learn to:

- ◆ Use the base multiplier property to write an exponential function.
- ◆ Solve an exponential equation for the base.
- ◆ Use two points to write an exponential function.

A. USING THE BASE MULTIPLIER TO FIND EXPONENTIAL FUNCTIONS

Recall from section 2.3 that for an exponential function of form $f(x) = ab^x$, if the value of the independent variable increases by 1, then the value of the dependent variable is multiplied by the base b. One way to find an equation for an exponential function of the form $y = ab^x$ is to use the base multiplier property.

▶ *Example 1*

Finding an Exponential Function from a Table of Values

An exponential function contains the points listed in the table in Figure 1. Write an equation to represent this function.

x	f (x)
0	4
1	12
2	36
3	108
4	324

Figure 1.

Solution

We recall from Section 2.3 that for the form $f(x) = ab^x$, a represents the initial value. Graphically this is the y-intercept $(0, a)$. From the table in Figure 1, then, we can see that the point $(0, 4)$ is on the curve, so $a = 4$. As the value of x increases by 1, the value of y is multiplied by 3. So the base multiplier $b = 3$. Putting these pieces together we have an equation for our function: $f(x) = 4(3)^x$.

For the next example, recall that for linear functions of the form $y = mx + b$, as x increases by 1, the value of y increases by the slope m. We can differentiate between linear and exponential functions by observing the behavior of the dependent variable.

▶ *Example 2*

Linear versus Exponential Functions

The tables in Figure 2 show points on the graphs of two functions. One is a linear function and one is an exponential function. Find an equation for each.

x	f (x)
0	−7
1	−1
2	5
3	11
4	17

x	g(x)
0	1280
1	320
2	80
3	20
4	5

Figure 2.

Solution

For f, as the value of x increases by 1, the value of y increases by 6. This indicates that f is a linear function with slope $m = 6$. The y-intercept is at $(0, -7)$. We can represent this function with the equation $f(x) = 6x - 7$.

For g, as the value of x increases by 1, the value of y divides by 4, which is equivalent to multiplying by $\frac{1}{4}$. This indicates that g is an exponential function with base multiplier $b = \frac{1}{4}$. The y-intercept is $(0, 1280)$. We represent this function with the equation $g(x) = 1280 \left(\frac{1}{4}\right)^x$.

Next we can use a graphing calculator and the table feature to check our work. Type each equation in the Y= screen to check. See Figure 3.

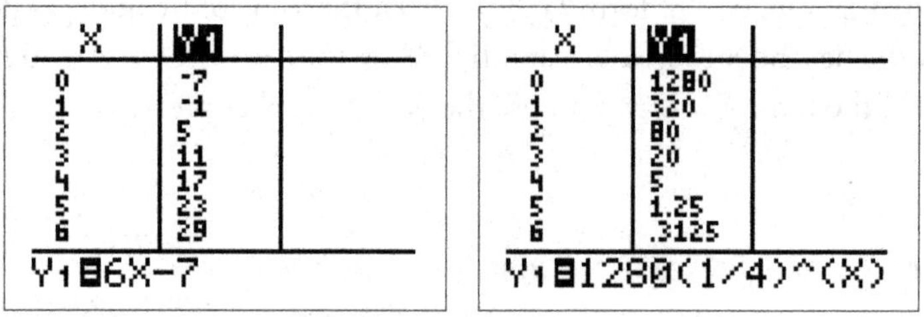

Figure 3. Verify equations with a table.

Practice Set A

The tables in Figure 4 show points on the graphs of two functions. Follow the instructions below, and then turn the page to check your work.

1. Find an equation for $f(x)$.

x	f (x)
0	500
1	100
2	20
3	4
4	0.8

2. Find an equation for $g(x)$.

x	g(x)
0	30
1	22
2	14
3	6
4	−2

Figure 4.

B. SOLVING EXPONENTIAL FUNCTIONS FOR THE BASE

Sometimes we are asked to write the equation for an exponential function but may not have a table of values with the x-values conveniently increasing by 1. If we are given just two points for a function, however, we can write a formula rule if we know how to solve equations of the form $ab^n = k$ for the base b.

▶ *Example 3*

Solving Equations for the Base

Find all real-number solutions to the following equations.

1. $b^2 = 144$

2. $b^3 = 200$

3. $12b^3 - 50 = -146$

4. $10b^4 = 6250$

5. $b^4 = -81$

6. $3b^6 = 17.5$

Solution

1. With this equation, $b = 12$ or $b = -12$ because $12^2 = 144$ and $(-12)^2 = 144$. We can say $b = \pm 12$.

2. Since there is no integer whose cube is 200, we use the properties of exponents to solve.

$$b^3 = 200 \qquad \text{Original equation.}$$
$$b = 200^{1/3} \qquad 200^{1/3} \text{ is the number whose third power is 200.}$$
$$b \approx 5.848 \qquad \text{Use a calculator and round to 3 decimal places.}$$

3. This equation requires two solving steps before we deal with the exponent.

$$12b^3 - 50 = -146 \qquad \text{Original equation.}$$
$$12b^3 = -96 \qquad \text{Add 50 to both sides.}$$
$$b^3 = -8 \qquad \text{Divide both sides by 12.}$$
$$(b^3)^{1/3} = (-8)^{1/3} \qquad (-8)^{1/3} \text{ is the number whose third power is } -8.$$
$$b = -2 \qquad (-2)^3 = -8$$

4. The exponent is even, so we will have two solutions, one positive and one negative:

$$10b^4 = 6250 \qquad \text{Original equation.}$$
$$b^4 = 625 \qquad \text{Divide both sides by 10.}$$
$$b = \pm 5 \qquad 5^4 = 625 \text{ and } (-5)^4 = 625$$

5. The equation $b^4 = -81$ has no real-number solutions because an even-numbered exponent always gives a positive solution.

6. Notice that our solving process can extend to any natural number n. Again, the exponent is even and equals a positive value, so we will have two solutions:

$3b^6 = 17.5$	Original equation.
$b^6 = 5.8\overline{3}$	Divide both sides by 3.
$(b^6)^{1/6} = 5.8\overline{3}^{1/6}$	$5.8\overline{3}^{1/6}$ is the number whose 6th power is $5.8\overline{3}$.
$b \approx \pm 1.342$	$1.342^{1/6} \approx 5.8\overline{3}$ and $(-1.342)^{1/6} \approx 5.8\overline{3}$.

We generalize the patterns we see in the previous example to outline the steps for solving equations of the form $b^n = k$ for b:

◆ If n is odd, the real-number solution is $k^{1/n}$.

◆ If n is even and $k \geq 0$, the real-number solutions are $\pm k^{1/n}$.

◆ If n is even and $k < 0$, there is no real-number solution.

Practice Set B

Find all real-numbered solutions. Round any results to three decimal places. When you're finished, you are welcome to turn the page and check your work.

3. $-2b^3 = -128$

4. $b^5 - 19 = 13$

5. $-5b^4 + 11 = -1.2$

6. $b^6 + 37 = 25$

Practice Set A — Answers

1. The table for $f(x)$ indicates an exponential function with base multiplier $b = \frac{1}{5}$ and y-intercept $(0, 500)$. $f(x) = 500\left(\frac{1}{5}\right)^x$.

2. The table for $g(x)$ indicates a linear function with slope $m = -8$ and y-intercept $(0, 30)$. $g(x) = -8x + 30$.

C. USING TWO POINTS TO FIND EXPONENTIAL FUNCTIONS

Now that we know how to solve an exponential equation for the base, we can look at another way to write equations for exponential curves. If we know two points on the curve, and if one of them is the y-intercept, $(0, a)$, then a is the initial value. Using a, we substitute the second point into the equation $y = ab^x$ and solve for b.

▶ *Example 4*

Finding an Equation of an Exponential Function

Find an equation of the form $y = ab^x$ that contains the two points given. If necessary, round the value of b to three decimal places.

   1. (0, 5) and (4, 405)   **2.** (0, 200) and (8, 86)

Solution

1. Since the y-intercept is $(0, 5)$, we know that $a = 5$ and that the equation has the form $y = 5b^x$. Next, we substitute the point $(4, 405)$ in the equation and solve for b.

$y = 5b^x$	The equation with $a = 5$.
$405 = 5b^4$	Substitute 4 for x and 405 for y.
$81 = b^4$	Divide both sides by 5.
$b = \pm 3$	The solutions to $b^4 = k$ are $b = \pm k^{1/4}$.

The base of an exponential function is positive, so we use the solution $b = 3$ to write the equation $y = 5(3)^x$. This equation models exponential growth.

We can now use a graphing calculator and the $\boxed{\text{TRACE}}$ button to verify our work, as in Figure 5.

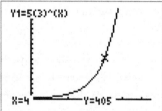

2. Because the y-intercept is $(0, 200)$, we know that $a = 200$ and that the equation has the form $y = 200b^x$. Next, we substitute the point $(8, 86)$ in the equation and solve for b.

Figure 5. Verify the equation.

$y = 200b^x$	The equation with $a = 200$.
$86 = 200b^8$	Substitute 8 for x and 86 for y.
$.43 = b^8$	Divide both sides by 200.
$b \approx \pm 0.900$	Solve for b. The solutions to $b^8 = k$ are $b = \pm k^{1/8}$.

Again we use only the positive solution $b = 0.9$ to write the equation $y = 200(0.9)^x$. This equation models exponential decay. You can verify the equation on your graphing calculator.

We can apply the process of finding an equation for an exponential curve to many real-life situations.

▶ *Example 5*

Writing an Exponential Model When the Initial Value Is Known

In 2006, 80 deer were introduced into a wildlife refuge. By 2012, the population had grown to 180 deer. The population was growing exponentially. Write a function, $N = f(t)$, which represents the population of deer over time, t. Assuming the trend continues, use your model to estimate the number of deer in the wildlife refuge in 2020.

Solution

We let our independent variable t be the number of years after 2006. Thus, the information given in the problem can be written as input-output pairs: (0, 80) and (6, 180). Notice that by choosing our input variable to be measured as years after 2006, we have given ourselves the initial value for the function, $a = 80$. We can now substitute the second point into the equation $N(t) = 80b^t$ to find b:

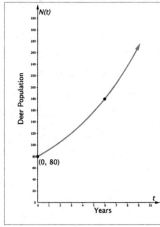

$$N(t) = 80b^t$$

$$180 = 80b^6 \qquad \text{Substitute using the point (6, 180).}$$

$$2.25 = b^6 \qquad \text{Divide both sides by 80.}$$

$$b = 2.25^{1/6} \qquad \text{The solution to } b^6 = k \text{ are } b = \pm k^{1/6}.$$

$$b \approx 1.1447 \qquad \text{Only use the positive solution. Round to 4 decimal places.}$$

The exponential model for the population of deer is

$$N(t) = 80(1.1447)^t.$$

To predict the population in 2020, let $t = 2020 - 2006 = 14$. Then,

$$N(14) = 80(1.1447)^{14} \approx 530.6$$

We predict that the population of deer will be about 531 in the year 2020. However, this exponential function may only model short-term growth. As the input becomes larger, the output will get increasingly larger — so much so that the model may not be useful in the long term.

We can graph our model to observe the population growth of deer in the refuge over time. Notice that the graph in Figure 7 passes through both points given in the problem, (0, 80) and (6, 180).

Figure 7. Graph of the deer population t years after 2006.

Practice Set B — Answers

3. $b = (-64)^{1/3} = -4$

4. $b = 32^{1/5} = 2$

5. $b = 2.44^{1/4} \rightarrow b \approx \pm 1.250$

6. $b^6 = -12$. There are no real-numbered solutions.

Practice Set C

Using the information given, write an exponential function of the form $y = ab^x$. If necessary, round the value of b to three decimal places. To check your solutions, turn the page.

7. Points on the curve (0, 3) and (6, 192)

8. Points on the curve (0, 180) and (5, 21)

9. The value of new online company is growing exponentially. In 2010, the company was worth 52 million dollars. By 2015, the value of the company is 96 million dollars. Let y be the value of the company in millions of dollars x years after 2010.

Exercises

For each table of values, determine if it represents a linear function or an exponential function. Then write an equation for the function. For a linear function use the form $f(x) = mx + b$. For an exponential function, use $f(x) = ab^x$. You should check your solutions using the table feature on your graphing calculator.

1.

x	$f(x)$
0	8
1	16
2	32
3	64
4	128

2.

x	$f(x)$
0	448
1	224
2	112
3	56
4	28

3.

x	$f(x)$
0	11
1	4
2	−3
3	−10
4	−17

4.

x	$f(x)$
0	2
1	10
2	50
3	250
4	1250

5.

x	$f(x)$
0	243
1	81
2	27
3	9
4	3

6.

x	$f(x)$
0	−9
1	1
2	11
3	21
4	31

Find all real-number solutions. Round your result(s) to three decimal places.

7. $b^2 = 49$

8. $b^4 = 16$

9. $b^5 = 243$

10. $b^3 = 216$

11. $-1000\,b^3 = 125$

12. $3b^6 = 340$

13. $12.3b^4 - 65.2 = 557$

14. $-20b^5 - 14.1 = 27.4$

15. $b^6 - 28.5 = -40.5$

16. $94b^2 + 61 = 5822$

17. $3.9b^3 + 9.1 = 11.4$

18. $6b^4 + 45 = 21$

For the following exercises, find the formula for an exponential function of the form $y = ab^x$ that passes through the given points. Round the value of b to three decimal places. Verify your equation with a graphing calculator.

19. (0, 6) and (3, 750)

22. (0, 14) and (5, 629)

25. (0, 11.77) and (8, 1.18)

20. (0, 2000) and (2, 20)

23. (0, 4.5) and (5, 11.2)

26. (0, 93.67) and (6, 5.24)

21. (0, 450) and (4, 18)

24. (0, 2.9) and (3, 11.2)

The following exercises give verbal descriptions of exponential models. For each,

   a. Use the information to write two points on the curve.

   b. Find an equation of the curve in the form $f(t) = ab^t$.

   c. Use your exponential model to make the prediction asked for.

27. Research biologists released 300 chicken turtles (*deirochelys reticularia*) into a wetland with hopes of repopulating the habitat. After 7 years, the research team estimated that the population had grown to 550 chicken turtles. Predict the population of chicken turtles in the habitat 12 years after the initial release.

28. Bacteria are growing in a Petri dish in a biology lab such that the original population of 200 bacteria have grown to 7800 bacteria in just 4 hours. Predict the number of bacteria after 6 hours.

29. In 2007, Linda purchased a new Toyota Prius priced at $19,500. In 2016, the Kelly Blue Book value of her car was just $5600. Estimate the value of her car in 2020.

30. In 2010, the value of a share of certain stock on the NASDAQ was $97.50, but the price per share of that stock has been decreasing each year so that the stock was only valued at $73.40 in 2016. Predict the value of the stock in 2025.

Practice Set C — Answers

7. $a = 3$, $b = 2$, and $y = 3(2)^x$

8. $a = 180$, $b \approx 0.651$, and $y = 180(0.651)^x$

9. We can write two points: (0, 52) and (5, 96). $a = 52$, $b \approx 1.130$, and $y = 52(1.13)^x$

2.5 Using Exponential Functions to Model Data

OVERVIEW

There are a multitude of situations that can be modeled by exponential functions, such as investment growth, radioactive decay, atmospheric pressure changes, and the temperature of a cooling object. We will take what we have learned in Section 2.4 about writing exponential equations and apply those skills to real world situations. You will learn to:

- Know the meaning of the base of an exponential model.
- For a model $f(t) = ab^t$, interpret the meaning of a and b in the context of the problem.
- Find an equation of an exponential model by using data from real world situations.
- Model a half-life situation.
- Use linear regression to find an exponential model for a set of data.
- Make estimates and predictions using an exponential model.

A. USING THE BASE MULTIPLIER TO FIND A MODEL

We continue our study of exponential models with a closer look at the meaning of the base multiplier, b, in equations of the form $f(t) = ab^t$. We want to understand exactly how the value of b is related to the percent rate of growth or decay over time. So let's take a second look at exponential growth and decay.

Exponential growth refers to an increase based on a constant multiplicative rate of change over equal increments of time. It is a constant *percent increase* of the original amount over time.

Exponential decay refers to a decrease based on a constant multiplicative rate of change over equal increments of time. It is a constant *percent decrease* of the original amount over time.

Percent change refers to a change based on a percent of the original amount. An exponential growth model will have the percent change — in decimal form — *added to 1* to indicate a percent increase of the original amount. An exponential decay model will have the percent change — in decimal form — *subtracted from 1* to indicate a percent decrease of the original amount over time.

▶ *Example 1*

Finding the Percent Change of an Exponential Model

State whether the functions below represent exponential growth or decay of a quantity over time. Then determine the percent change per unit of time.

1. $P(t) = 45(1.07)^t$

2. $f(t) = 6600(0.92)^t$

3. $g(t) = 812(0.75)^t$

4. $A(t) = 3000(1.038)^t$

5. $P(t) = 3(2)^t$

Solution

1. $b = 1.07$, so $b > 1$, indicating exponential growth. $1.07 - 1 = 0.07$, which indicates a 7% increase per unit of time.

2. $b = 0.92$, so $0 < b < 1$, indicating exponential decay. $1 - 0.92 = 0.08$, which indicates an 8% decrease per unit of time. Although there's an 8% decrease over time, this number is not apparent in the equation. What we see is $b = 0.92$, which indicates that although we *lose* 8% over time, what *remains* is the other 92%.

3. $b = 0.75$, so $0 < b < 1$, indicating exponential decay. $1 - 0.75 = 0.25$, which indicates a 25% decrease per unit of time. 75% remains.

4. $b = 1.038$, so $b > 1$, indicating exponential growth. $1.038 - 1 = 0.038$, which indicates a 3.8% increase per unit of time.

5. $b = 2$, so $b > 1$, indicating exponential growth. $2 - 1 = 1$, which indicates a 100% increase per unit of time.

> ### Percent Rate of Change of an Exponential Model
>
> If $f(t) = ab^t$, where $a > 0$, models a quantity at time t, then the percent rate of change is constant. Specifically, for a percent in decimal form,
>
> - If $b > 1$, then the quantity grows exponentially at a rate of $b - 1$ percent per unit of time.
>
> - If $0 < b < 1$, then the quantity decays exponentially at a rate of $1 - b$ percent per unit of time.

The last equation from Example 1, $P(t) = 3(2)^t$, shows a 100% increase, which means that 100% is added to the quantity already present. In essence, 100% of a quantity has grown to 200% of that quantity. It is easier to think of this as "doubling." In fact, an exponential model with base $b = 2$ is often referred to as a "doubling function." Likewise, if the base $b = 3$, we say a quantity "triples" over time .

C. FINDING A MODEL USING DATA DESCRIBED IN WORDS

We can use the concept of a constant percent change with the initial amount to write and interpret exponential equations. For example, in Section 2.4, we found the exponential model for a population of deer over t years: $N(t) = 80(1.1447)^t$. For this model, when $t = 0$, $N(0) = 80$, which means that the initial quantity of deer was 80. The base $b = 1.1447 = 1 + 0.1447$, indicates a 14.47% increase in the population per year.

> ### Exponential Models
>
> If $f(t) = ab^t$, where $a > 0$, models a quantity at time t, then
>
> - a represents the initial quantity at time $t = 0$.
>
> - b represents $1 \pm r$, where r is the percent increase or decrease of the quantity over time.

▶ Example 2

Modeling with an Exponential Function

Write an exponential function of the form $f(t) = ab^t$ to represent the situation described.

1. The initial value is $4545, increasing 6% annually.

2. The original amount is 132 grams, decaying 7.6% every hour.

3. The beginning population is 2.7 million, growing 1.89% annually.

4. The initial quantity is 250 milligrams, decaying 50% every 1000 years.

5. An initial stock investment of $5000 doubles every 8 years.

Solution

1. $a = 4545$ and $b = 1 + 0.06 = 1.06$, so $f(t) = 4545(1.06)^t$, where t is measured in years.

2. $a = 132$ and $b = 1 - 0.076 = 0.924$, so $f(t) = 132(0.924)^t$, where t is measured in hours.

3. $a = 2.7$ and $b = 1 + 0.0189 = 1.0189$, so $f(t) = 2.7(1.0189)^t$, where t is measured in years.

4. $a = 250$ and $b = 1 - 0.50 = 0.5$, so $f(t) = 250(0.5)^{t/1000}$, where t is measured in years. Notice that the t variable in this model is divided by 1000. We can rewrite this equation using exponent rules:

$$f(t) = 250(0.5)^{t/1000} = 250((0.5)^{1/1000})^t \approx 250(.9993)^t$$

Using the second form of the model, we say that the quantity decreases by 0.0007 % per year rather than by 50% every 1000 years.

5. $a = 5000$ and $b = 1 + 1.0 = 2$, so $f(t) = 5000(2)^{t/8}$, where t is measured in years. As in the previous problem, we can also rewrite this equation:

$$f(t) = 5000(2)^{t/8} = 5000((2)^{1/8})^t = 5000(1.0905)^t$$

The second form of the model indicates an increase of 9.05% per year.

Among the applications of exponential functions are situations involving interest-bearing accounts. An initial investment that earns r percent interest is compounded annually. In this situation, the interest earned each year equals r percent of the principal *and* r percent of any interest earned in the previous years.

▶ *Example 3*

Modeling Interest Compounded Annually

A person invests $3000 in an account that earns 4.5% interest compounded annually.

a. Let $f(t)$ be the value (in dollars) of the investment after t years. Write an equation for f.

b. What will be the value of the investment after 5 years? After 10 years?

Solution

a. We use the model for exponential growth $f(t) = ab^t$. The initial value $a = 3000$. The base b represents a 4.5% increase, so $b = 1 + .045 = 1.045$.

$$f(t) = 3000(1.045)^t$$

b. To find the value in 5 years and in 10 years, we let $t = 5$ and then $t = 10$, so:

$$f(5) = 3000(1.045)^5 \approx 3738.55$$
$$f(10) = 3000(1.045)^{10} \approx 4658.91$$

The value of the investment will be $3738.55 in 5 years and $4658.91 in 10 years.

We can also represent the exponential function in Example 3 in table form and graphically. Table 1 presents the graphing calculator graph of $y = 3000(1.045)^x$. The y-intercept $(0, 3000)$ corresponds to the initial investment, and the function increases exponentially.

Figure 2 shows a table of the first years of the investment. Notice that after the first year, the investment earns interest on the principal *and* on the interest earned in previous year(s).

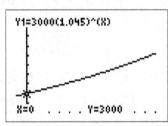

Figure 1.

t	$f(t)\ =\ 3000(1.045)^t$
0	$3000(1.045)^0 = 3000$
1	$3000(1.045)^1 = 3135$
2	$3135(1.045)^2 = 3276.08$
3	$3276.08(1.045)^3 = 3423.50$
4	$3423.50(1.045)^4 = 3577.56$
5	$3577.56(1.045)^5 = 3738.55$

Figure 2. Values of a compounded interest account.

Practice Set C

State whether the functions below represent exponential growth or decay. Then determine the percent change.

1. $g(t) = 400(0.88)\,t$ **2.** $P(t) = 9.9(1.35)^t$ **3.** $f(t) = 7000(1.022)^t$

A person invests \$2500 in an account that earns 6.33% interest compounded annually. Use this situation to answer questions 4 and 5.

4. Let $f(t)$ be the value (in dollars) of the investment after t years. Write an equation for f.

5. What will be the value of the investment after 4 years? After 8 years?

Turn the page to check your solutions.

D. EXPONENTIAL DECAY AND HALF-LIFE APPLICATIONS

We've seen that if the base of our exponential function is between 0 and 1, we have a model of exponential decay, a decreasing function.

▶ Example 4

Modeling Exponential Decay

A car tire that initially has air pressure of 33 psi (pounds per square inch) has begun leaking. The rate at which pressure in the tire is lost is about 4% per minute.

a. Let $f(t)$ represent the air pressure in the tire (in psi) t minutes after it begins leaking. Write an equation for f.

b. Predict the air pressure in the tire after 5 minutes and then after a half hour.

Solution

a. The initial amount $a = 33$. The base is found by subtracting the percent decrease from 1:

$b = 1 - 0.04 = 0.96$. If 4% of the air pressure is lost each minute, then 96% of the air pressure remains The equation is: $f(t) = 33(0.96)^t$

b. After 5 minutes, $t = 5$, and after a half hour, $t = 30$, so:

$$f(5) = 33(0.96)^5 \approx 26.9. \text{ After 5 minutes, the air pressure is 26.9 psi.}$$

$$f(30) = 33(0.96)^{30} \approx 9.7. \text{ After 30 minutes, the air pressure is 9.7 psi.}$$

We can also describe how quickly a quantity decays by its **half-life**. If a quantity decays exponentially, the half-life is the amount of time it takes for that quantity to be reduced to half its initial value. The term is commonly used in physics to describe how quickly unstable atoms undergo radioactive decay. The medical sciences often refer to the biological half-life of drugs and other chemicals in the body.

For example, carbon-14 is a radioactive isotope of carbon that is often used to date organic remains from archeological sites. The half-life of carbon-14 is about 5730 years, which means that 100 grams of carbon-14 would decay to just 50 grams of carbon-14 over a period of 5730 years.

▶ Example 5

Modeling Radioactive Decay

An archaeologist estimates that the initial amount of carbon-14 in a fossil sample was 92 milligrams. The half-life of carbon-14 is about 5730 years.

    a. Write an equation for $f(t)$ representing the amount of carbon-14 (in mg) remaining in the fossil after t years.

    b. Use your model to estimate the amount of carbon-14 remaining in the fossil after 20,000 years.

Solution

a. The initial value $a = 92$. The base represents a 50% decrease, which is equivalent to 50% of the initial amount remaining. So $b = 1 - 0.50 = 0.50$. However, it is conventional to write the base as a fraction, $b = \frac{1}{2}$, when modeling a half-life application. The equation is:

$$f(t) = 92\left(\frac{1}{2}\right)^{t/5730}$$

b. For $t = 20000$, $f(20000) = 92\left(\frac{1}{2}\right)^{\frac{20000}{5730}} \approx 8.19$. After 20,000 years, there will be about 8.19 mg of carbon-14 remaining in the fossil.

When calculating the output of an exponential function with a fractional exponent, we may need to put parentheses around the exponent — depending on the calculator model. Figure 3 presents ways to enter this calculation.

```
92(1/2)^(20000/5730)        92(1/2)^(20000/5
          8.186010742       730)
                                      8.186010742
                            92*.5^(20000/573
                            0)
                                      8.186010742
```

Figure 3. Computing the output of a half-life function.

Practice Set D

Now it's your turn to work with real-world models. When you are finished, turn the page to check your solutions.

6. A new computer that cost $2200 depreciates in value by about 13% per year. Write an exponential equation that models the value of the computer (in dollars) after t years. Use the model to estimate the value after 6 years.

7. The half-life of caffeine in a person's bloodstream is about 6 hours. Write an exponential functions that models the amount of caffeine (in mg) remaining in a person's bloodstream after t hours if they began with 100 milligrams of caffeine. Use the model to predict the amount of caffeine remaining after 8 hours.

E. FINDING A MODEL USING DATA IN A TABLE AND EXPONENTIAL REGRESSION

In Section 2.4, we found equations for exponential functions using two points when one of those points was the y-intercept $(0, a)$. In this book, if we have several data points, or if we do not know the initial value, we will use **exponential regression** on a graphing calculator to find a model for our data.

To do that, we enter the data into Lists 1 and 2 and use the command ExpReg on the calculator to fit an exponential function to a set of data points. This returns an equation of the form $y = ab^x$.

From this equation, we know the following:

♦ The initial value of the model is $y = a$.

♦ If $b > 1$, the function models exponential growth.

♦ If $0 < b < 1$, the function models exponential decay.

These are the steps you'll need to follow to perform exponential regression using a graphing calculator:

1. Use the STAT then Edit menu to enter given data. Clear any existing data from the lists. List the input values in the L1 column. List the output values in the L2 column.

2. Graph and observe a scatter plot of the data using the [STAT PLOT] feature. Use ZOOM [9] to adjust axes to fit the data. Then verify that the data follows an exponential pattern.

3. Find the equation that models the data. Select ExpReg from the STAT then [CALC] menu. Use the values returned for a and b to record the model, $y = ab^x$.

4. Graph the model in the same window as the scatterplot to verify it is a good fit for the data.

Practice Set C — Answers

1. $b = 0.88$, indicating exponential decay.
$1 - 0.88 = 0.12$, a 12% decrease over time.

2. $b = 1.35$, indicating exponential growth.
$1.35 - 1 = 0.35$, a 35% increase over time.

3. $b = 1.022$, indicating exponential growth.
$1.022 - 1 = 0.022$, a 2.2% increase over time.

4. $f(t) = 2500(1.0633)^t$

5. $f(4) = 2500(1.0633)^4 \approx 3195.68$ and $f(8) = 2500(1.0633)^8 \approx 4084.95$ The value of the investment is thus $3195.68 after 4 years and $4084.95 after 8 years.

► *Example 6*

Using Exponential Regression to Fit a Model to Data

In 2007, Indiana University published a study investigating the crash risk of alcohol-impaired driving. The study used data from 2,871 crashes to measure the association of a person's blood alcohol level (BAC) with the risk of being in an accident. Figure 4 presents some results from the study. The relative risk is a measure of how many times more likely a person is to crash. So, for example, a person with a BAC of 0.09 is 3.54 times as likely to crash as a person who has not been drinking alcohol.

BAC	0	0.01	0.03	0.05	0.07	0.09	0.11	0.13	0.15	0.17	0.19	0.21
Relative Risk of Crashing	1	1.03	1.06	1.38	2.09	3.54	6.41	12.6	22.1	39.05	65.32	99.78

Figure 4. The crash risk of alcohol-impaired driving. From Indiana University Center for Studies of Law in Action (2007).

Use the data from Figure 4 to do the following.

a. Let x represent the BAC level, and let y represent the corresponding relative risk. Use exponential regression to fit a model to these data.

b. After 6 drinks, a person weighing 160 pounds will have a BAC of about 0.16. How many times more likely is this person to crash if they drive after having a 6-pack of beer?

Solution

a. Using the [STAT] and then the Edit menu on a graphing utility, list the BAC values in L1 and the relative risk values in L2. Then use the [STAT PLOT] feature to verify that the scatter plot follows an exponential pattern. See Figure 5.

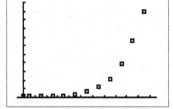

Figure 5. Use Zoom Fit to see the scatter plot.

Next, we use the ExpReg command from the [STAT] and then [CALC] menus to obtain the exponential model, $y = 0.58304829\,(2.20720213\text{E}10)^x$. See Figure 6.

Converting the base from scientific notation, we have $y = 0.58304829\,(22{,}072{,}021{,}300)^x$. We can then graph the model in the same window as the scatter plot to verify it is a good fit, as shown in Figure 7.

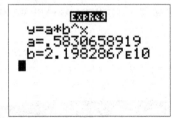

Figure 6. Use the ExpReg command.

b. To estimate the risk associated with a BAC of 0.16, we substitute 0.16 for x in the model and calculate y.

$$y = 0.58304829\,(22{,}072{,}021{,}300)^x$$
$$= 0.58304829\,(22{,}072{,}021{,}300)^{0.16}$$
$$\approx 26.35$$

If a 160-pound person drives after having 6 drinks, he or she is about 26.35 times more likely to crash than if driving while sober.

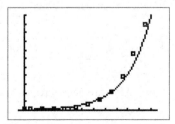

Figure 7. The regression model and scatter plot show a good fit.

▶ *Example 7*

Interpreting an Exponential Regression Model

The data in Figure 9 shows the world population in billions according to the U.S. Census Bureau over the last century.

For this problem, let $f(x)$ be the world population (in billions) x years after 1900, so that $x = 30$ corresponds to the year 1930, $x = 100$ corresponds to the year 2000, and so on.

a.   Use exponential regression to fit a model to this data.

b.   Interpret the meaning of the base b in the context of the problem.

c.   Interpret the meaning of the y-intercept in the context of the problem.

d.   Evaluate and interpret $f(125)$.

Year	Population (billions)
1930	2.070
1940	2.295
1950	2.500
1960	3.050
1970	3.700
1980	4.454
1990	5.279
2000	6.080
2010	6.866

Figure 9. World population figures according to the U.S. Census Bureau.

Solution

a. Rounding the values of a and b to three significant digits, the regression model is $y = f(x) = 1.210(1.0161)^x$.

b. The model represents exponential growth because $b > 1$. More specifically, $b = 1.0161 = 1 + 0.0161$, which tells us that the world population is increasing by 1.61% per year.

c. The y-intercept is $(0, 1.210)$ and indicates that the world population in 1900, the initial year in this problem, was about 1.210 billion.

d. $y = f(125) = 1.210(1.0161)^{125} \approx 8.909$

Our model predicts the world population will be about 8.909 billion in 2025. Notice that if you use the regression equation without rounding a and b, the prediction is 8.926 billion people. This discrepancy gives you an idea about just how sensitive a model can be when it comes to rounding.

Practice Set E

Figure 10 shows a recent graduate's credit card balance each month after graduation. Use this data to solve the following problems. When you are finished, turn the page to check your solutions.

Month	1	2	3	4	5	6	7	8
Debt ($)	620.00	761.88	899.80	1039.93	1270.63	1589.04	1851.31	2154.92

Figure 10.

8.   Use exponential regression to fit a model to these data.

9.   Determine the percent increase of the model.

10. If spending continues at this rate, what will the graduate's credit card debt be one year after graduating?

Exercises

State whether the functions below represent exponential growth or decay of a quantity over time. Then determine the percent change per unit of time.

1. $f(t) = 225(1.19)^t$

2. $P(t) = 4.1(1.64)^t$

3. $g(t) = 2000(0.98)^t$

4. $h(t) = 7.48(0.87)^t$

5. $h(t) = 9.13(1.0285)^t$

6. $f(t) = 290(1.0854)^t$

7. $A(t) = 300\left(\frac{1}{2}\right)^{t/12}$

8. $Q(t) = 12\left(\frac{1}{2}\right)^{t/5}$

Write an exponential function of the form $f(t) = ab^t$ to represent the situation described. Assume that t is measured in years.

9. The beginning population is 42,000, growing 2.54% annually.

10. The beginning population is 1.6 million, decreasing 4.1% annually.

11. The initial quantity is 250 milligrams, decaying 50% every 28 years.

12. The initial value is $7200, increasing 5% annually

13. A stock investment of $1500 doubles every 12 years.

14. The original amount is 16 grams, decaying 7.6% every hour.

15. The initial value is $2340, depreciating by 9.5% per year.

16. An investment of $450 triples every 22 years.

17. A person invests $2700 in an account that earns 3.25% interest compounded annually.

    a. Let $f(t)$ be the value (in dollars) of the account after t years. Write an equation for f.

    b. What will be the value of the account after 5 years? After 10 years?

18. A person invests $2700 in an account that earns 7.1% interest compounded annually.

    a. Let $f(t)$ be the value (in dollars) of the account after t years. Write an equation for f.

    b. What will be the value of the investment after 6 years? After 12 years?

19. Someone invests $1000 in stocks today, and the stocks' value doubles every 7 years.

    a. Let $f(t)$ be the value (in dollars) of the investment after t years. Write an equation for f.

    b. What will be the value of the investment 25 years from now?

20. Someone invests $9000 in stocks today and the stocks' value doubles every 12 years.

    a. Let $f(t)$ be the value (in dollars) of the investment after t years. Write an equation for f.

    b. What will be the value of the investment after 5 years? After 10 years?

21. A small coastal fishing town in Oregon has been losing job opportunities and losing population since 1990. In 1990, the population was 9740 but it has been decreasing by about 7.2% per year.

    a. Write an exponential function P that models the population t years since 1990.

    b. If this trend continues, predict the population in 2025.

Practice Set D — Answers

6. $a = 2200$, $b = 1 - 0.13 = 0.87$, and
    $f(t) = 2200(0.87)^t$
    $f(6) = 2200(0.87)^6 \approx 953.98$, so The computer is worth about $954 after 6 years.

7. $a = 100$, $b = \frac{1}{2}$, and $f(t) = 100\left(\frac{1}{2}\right)^{t/6}$
    $f(8) = 100\left(\frac{1}{2}\right)^{8/6} \approx 39.7$, so after 8 hours, there are 39.7 mg of caffeine in the bloodstream.

22. A small logging town in Oregon has been losing job opportunities and losing population since 2000. In 2000, the population was 13,400 but it has been decreasing by about 8.1% per year.

    a. Write an exponential function that models the population t years since 2000.

    b. If this trend continues, predict the population in 2020.

23. A sport utility vehicle (SUV) that cost $32,000 new depreciates in value by about 24% per year after it is purchased. Write an equation to model the value V of the SUV t years after it is purchased and use the model to predict the value of the car after 4 years.

24. The value of a new printing press is $375,000. This value is expected to depreciate by about 22% per year for the first 7 years of operation. Write an equation to model the value V of the press after t years of operation and use the model to predict the value of the press after 7 years.

25. The half-life of radioactive radium (Ra-226) is 1620 years. Write an equation to model the amount of Ra-226 left after t years in a sample that originally contained 40 mg of the isotope. Use the model to estimate the amount of Ra-226 remaining after 2000 years.

26. An archaeologist estimates that the initial amount of carbon-14 in a fossil sample was 26 milligrams. The half-life of carbon-14 is about 5730 years. Write an equation to model the amount of carbon-14 (in mg) remaining in the fossil after t years. Use the model to estimate the amount of carbon-14 remaining after 12,000 years.

27. The half-life of aspirin in a person's bloodstream is about 20 minutes. If a person's bloodstream contains 200 milligrams of aspirin, how much of that aspirin will remain after 2 hours?

28. The half-life of caffeine in a person's bloodstream is about 6 hours. A person who has just consumed an energy drink has about 242 milligrams of caffeine in their bloodstream. How much caffeine remains after 5 hours?

Use a graphing calculator for the remaining problems.

29. Figure 11 shows the population in the U.S. for selected years over the last century.

    a. Let y be the U.S. population in millions x years since 1900. Create a scatter plot of the data and calculate an exponential regression model. Round values to three decimal places.

    b. Graph the regression equation in the same window as the scattergram.

    c. Determine the approximate percent change of the population over time.

    d. Predict the population of the U.S. in 2025.

Year	US Population (millions)
1920	106.5
1940	132.1
1960	180.7
1980	226.5
2000	282.2
2014	318.9

Figure 11. U.S. population data from the U.S. Census Bureau.

Practice Set E — Answers

8. Rounding to three decimal places, the exponential regression model that fits these data is $y = 522.886(1.196)^x$.

9. $1.196 - 1 = 0.196$, so the percent increase is about 19.6% per month.

10. If spending continues at this rate, the graduate's credit card debt will be about $4,479 after one year ($x = 12$). Yikes.

30. Figure 12 shows the U.S. gross domestic product (GDP) in trillions of U.S. dollars (USD) for selected years.

    a.   Let *y* be the U.S. GDP in trillions of USD *x* years since 1900. Create a scatter plot of the data and calculate an exponential regression model. Round values to three decimal places.

    b.   Graph the regression equation in the same window as the scattergram.

    c.   Determine the approximate percent change of the GDP over time.

    d.   Predict the U.S. GDP in 2030.

Year	U.S. GDP (Trillion $)
1960	0.543
1970	1.076
1980	2.862
1990	5.980
2000	10.280
2010	14.960
2015	17.914

Figure 12. U.S. gross domestic product.

31. Figure 13 shows the revenue in billions of dollars collected by the U.S. Internal Revenue Service (IRS) for selected years.

    a.   Let *y* be revenue collected by the IRS (in billions of dollars) *x* years since 1900. Create a scatter plot of the data and calculate an exponential regression model. Round values to three decimal places.

    b.   Graph the regression equation in the same window as the scattergram.

    c.   Interpret the meaning of the values of *a* and *b* in the context of the problem.

    d.   Estimate the revenue collected by the IRS in 2018.

Year	Revenue (Billion $)
1960	91.8
1965	114.4
1970	195.7
1975	293.8
1980	519.4
1985	742.9
1990	1056.4
1995	1375.7
2000	2096.9

Figure 13. Revenue collected by the U.S. Internal Revenue Service.

Year	Number of Screens (thousands)
1975	11
1980	14
1985	18
1990	23
1995	27
2000	37

Figure 14. Number of movie theater screens in the U.S.

32. Figure 14 shows the number of movie theater screens (in thousands) in the U.S. for selected years.

    a.   Let *y* be number of movie screens (in thousands) *x* years since 1975. Create a scatter plot of the data and calculate an exponential regression model. Round values to three decimal places.

    b.   Graph the regression equation in the same window as the scattergram.

    c.   Interpret the meaning of the values of *a* and *b* in the context of the problem.

    d.   Predict the number of movie screens in 2018.

33. Figure 15 presents the number of computers connected to the Internet for selected years.

   a. Let y be the number of computers connected to the Internet x years since 1983. Create a scatter plot of the data and calculate an exponential regression model. Round values to three decimal places.

   b. Graph the regression equation in the same window as the scattergram.

   c. Estimate the number of computers on the Internet in 2000.

Year	Approximate Number Of Computers On The Internet
1983	562
1984	1,024
1985	1,961
1986	2,308
1987	5,089
1988	28,174
1989	80,000
1990	290,000
1991	500,000
1992	727,000
1993	1,200,000
1994	2,217,000

Figure 15. Approximate number of computers on the Internet.

CHAPTER 3

Logarithmic Functions

In Chapter 2, we studied exponential functions. In this chapter, we study logarithmic functions, which are related to exponential functions. Logarithmic functions are quite useful in helping us solve exponential equations for the variable, and that will be our main focus in Sections 3.2 and 3.3. We will also study some properties of logarithms and discuss how they relate to the properties of exponents.

In addition to being a sort of solving tool for exponential equations, logarithmic functions themselves have many real-world applications. One of the most familiar uses of logarithms is to measure the relative magnitude of earthquakes using the Richter scale. The Richter scale is a base-ten logarithmic scale. It describes how an earthquake of magnitude 8 is not twice as great as an earthquake of magnitude 4. It's *10,000 times* as great because $10^{8-4} = 10^4 = 10,000$. We will investigate the nature of the Richter scale and the base-ten logarithm upon which it is defined.

In this chapter, we'll cover the following topics:

3.1 Logarithmic Functions

OVERVIEW

In this section, we define a new function called a logarithm. This new function will generate exponents on a base to help us answer questions such as, "10 to what exponent equals 500?" This is equivalent to solving the equation $10^x = 500$ for x.

In this section, you will learn to:

- ◆ Know the meaning of logarithm and logarithmic function.
- ◆ Find logarithms.
- ◆ Evaluate logarithmic functions.
- ◆ Know the basic properties of logarithms.

A. DEFINITION OF LOGARITHM

Consider the function $y = f(x) = 2^x$. The table in Figure 1 shows some input-output pairs for f.

For this function, the input is the exponent on base 2, and the output is the power of 2. So if the input is $x = 3$, then 3 is the exponent on base 2, and the output is $y = 2^3 = 8$.

A **logarithm base** b, written $\log_b(x)$, is a function that inverts or exchanges the input-output pairs of an exponential function with the same base. For a logarithm, the input is the power of the base and the output is the exponent. Using our example, if $x = 8$ is the input to a logarithm with base 2, then the output is 3, because $2^3 = 8$. For $\log_2(x) = y$, then:

x	$y = 2^x$
0	$1 = 2^0$
1	$2 = 2^1$
2	$4 = 2^2$
3	$8 = 2^3$
4	$16 = 2^4$

Figure 1. Some input-output pairs for $y = 2^x$.

$$\log_2(1) = 0 \text{ because } 2^0 = 1$$

$$\log_2(2) = 1 \text{ because } 2^1 = 2$$

$$\log_2(4) = 2 \text{ because } 2^2 = 4$$

$$\log_2(8) = 3 \text{ because } 2^3 = 8$$

A logarithm base 2 gives back the exponent when the input is a power of 2. In fact, a logarithm *always* equals an exponent on the base.

Logarithm Base b

For $b > 0$, $b \neq 1$, and $x > 0$, the logarithm $\log_b(x)$ is the number y such that $b^y = x$.

In words, we say "log base b of x" is the exponent we raise b to in order to get x.

Evaluating Logarithms

Knowing the squares, cubes, and roots of numbers allows us to evaluate many logarithms mentally. Given a logarithm of the form $\log_b(x)$, we can evaluate it mentally by following these steps:

1.   Rewrite the input x as a power of b: $b^y = x$.

2.   Use previous knowledge of powers of b to identify y by asking, "To what exponent should b be raised in order to get x?"

For example, consider $\log_7(49)$. We ask, "To what exponent must 7 be raised in order to get 49?" We know $7^2 = 49$. Therefore, we know that $\log_7(49) = 2$.

▶ *Example 1*

Finding Logarithms Mentally

Find the logarithms.

1. $\log_3(81)$ 3. $\log_5(125)$ 5. $\log_{10}(1,000,000)$

2. $\log_8(64)$ 4. $\log_2(32)$ 6. $\log_9 1$

Solution

1. $\log_3(81) = 4$ because $3^4 = 81$

2. $\log_8(64) = 2$ because $8^2 = 64$

3. $\log_5(125) = 3$ because $5^3 = 125$

4. $\log_2(32) = 5$ because $2^5 = 32$

5. $\log_{10}(1,000,000) = 6$ because $10^6 = 1,000,000$

6. $\log_9(1) = 0$ because $9^0 = 1$

In order to tackle some trickier logarithms, let's recall the meaning of negative exponents. The general rule, $b^{-n} = \frac{1}{b^n}$, reminds us that if the power of the base is in the denominator of a fraction, it can also be written with a negative exponent. So, for example:

$$\log_9\left(\frac{1}{81}\right) = -2 \quad \text{because } 9^{-2} = \frac{1}{9^2} = \frac{1}{81}$$

▶ *Example 2*

Finding the Logarithm of a Reciprocal

Find the logarithm.

1. $\log_2\left(\frac{1}{16}\right)$ 2. $\log_4\left(\frac{1}{64}\right)$

Solution

1. $\log_2\left(\frac{1}{16}\right) = -4$ because $2^{-4} = \frac{1}{2^4} = \frac{1}{16}$

2. $\log_4\left(\frac{1}{64}\right) = -3$ because $4^{-3} = \frac{1}{4^3} = \frac{1}{64}$

Now recall the relationship between fractional exponents and roots:

$$b^{1/2} = \sqrt{b} \text{ and } b^{1/3} = \sqrt[3]{b}$$

To find $\log_{25} 5$, we ask, "25 to what exponent equals 5?" We see $\log_{25}(5) = \frac{1}{2}$ because $25^{1/2} = \sqrt{25} = 5$.

▶ **Example 3**

Finding Logarithms of Roots

Find the logarithm.

   1. $\log_8(2)$ **2.** $\log_{49}(7)$ **3.** $\log_{25}\left(\frac{1}{5}\right)$

Solution

1. $\log_8(2) = \frac{1}{3}$ because $8^{1/3} = \sqrt[3]{8} = 2$

2. $\log_{49}(7) = \frac{1}{2}$ because $49^{1/2} = \sqrt{49} = 7$

3. $\log_{25}\left(\frac{1}{5}\right) = -\frac{1}{2}$ because $25^{-1/2} = \frac{1}{25^{1/2}} = \frac{1}{\sqrt{25}} = \frac{1}{5}$

The equations in this last example incorporate both the rule for negative exponents and the definition of fractional exponents.

Practice Set A

Find the logarithm. When you are finished, turn the page to check your work.

  1. $\log_3(27)$ **3.** $\log_2(32)$ **5.** $\log_5\left(\frac{1}{125}\right)$

  2. $\log_6(36)$ **4.** $\log_4\left(\frac{1}{16}\right)$ **6.** $\log_7(\sqrt{7})$

B. COMMON LOGARITHMS

Sometimes you may see a logarithm written without a base. In this case, assume that the base is 10. In other words, the expression $\log(x)$ means $\log_{10}(x)$. A base-10 logarithm is called a **common logarithm**. The reason for the name "common" is that our numeration system, the one we all use in daily life, is structured on the base of 10. We read $\log(x)$ as, "the logarithm with base 10 of x" or just "log base 10 of x."

Common logarithms are used to define the Richter Scale, a system used to measure the relative strength of earthquakes. Scales for measuring the brightness of stars and the pH of acids and bases also use common logarithms.

The Common Logarithm

A **common logarithm** is a logarithm with base 10. We write $\log_{10}(x)$ simply as $\log(x)$. The common logarithm of a positive number x satisfies the following definition.

    If $x > 0$, $y = \log(x)$ is equivalent to $10^y = x$

    The logarithm y is the exponent to which 10 must be raised to get x.

Given a common logarithm of the form $y = \log(x)$, we can evaluate it mentally by following these steps:

1. Rewrite the input x as a power of 10: $10^y = x$.

2. Use previous knowledge of powers of 10 to identify y by asking, "To what exponent must 10 be raised in order to get x?"

Let's make the process a little bit easier by looking at Figure 2, a table of some of the powers of 10.

Exponent	Power of 10
-3	$10^{-3} = \frac{1}{10^3} = \frac{1}{1000} = 0.001$
-2	$10^{-2} = \frac{1}{10^2} = \frac{1}{100} = 0.01$
-1	$10^{-1} = \frac{1}{10^1} = \frac{1}{10} = 0.1$
0	$10^0 = 1$
1	$10^1 = 10$
2	$10^2 = 100$
3	$10^3 = 1000$
4	$10^4 = 10,000$
5	$10^5 = 100,000$
6	$10^6 = 1,000,000$

Figure 2. Some powers of 10.

▶ **Example 4**

Finding the Value of a Common Logarithm Mentally

Find the logarithm.

1. $\log(100,000)$
2. $\log(1,000,000,000)$
3. $\log(\frac{1}{10000})$
4. $\log(.01)$
5. $\log(1)$

Solution

1. $\log(100,000) = 5$ because $10^5 = 100,000$

2. $\log(1,000,000,000) = 9$ because $10^9 = 1,000,000,000$

3. $\log(\frac{1}{10000}) = -4$ because $10^{-4} = \frac{1}{10^4} = \frac{1}{10000}$

4. $\log(.01) = -2$ because $10^{-2} = \frac{1}{100} = .01$

5. $\log(1) = 0$ because $10^0 = 1$

In order to compare the magnitudes of two different earthquakes, we need to be able to convert between logarithmic and exponential form. For example, suppose the amount of energy released from one earthquake is 500 times greater than the amount of energy released from another. We want to calculate the difference in magnitude or the difference in the Richter numbers of the two quakes. Because a Richter number is a base-10 logarithm, the equation that represents this problem is $10^x = 500$, where x represents the difference in magnitudes on the Richter scale.

How would we solve for x?

We haven't yet learned a method for solving exponential equations. None of the algebraic tools discussed so far is sufficient to solve $10^x = 500$. But we do know that $10^2 = 100$ and $10^3 = 1000$, so it is clear that x must be some value between 2 and 3 because $y = 10^x$ is an increasing (exponential growth) function.

We can examine a graph, as in Figure 3, to better estimate the solution. We see that the value of x that satisfies $10^x = 500$ is between 2.5 and 3. Estimating from a graph, however, is still imprecise. To find an algebraic solution, we must use a common logarithm so that the input is the power on base 10, namely 500, and the output is the exponent.

We have $\log_{10}(500) = \log(500) = x$. By definition, this is equivalent to the equation $10^x = 500$. Now we can use a calculator to evaluate a common logarithm with the form $y = \log(x)$ by following these steps:

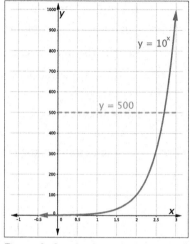

Figure 3. Graph of $y = 10^x$.

1. Press $\boxed{\text{LOG}}$.

2. Enter the value given for x, followed by $\boxed{)}$.

3. Press $\boxed{\text{ENTER}}$.

We enter $\log(500)$, and we see that the value is about 2.6990, as in Figure 4. This means that $x \approx 2.6990$ is a solution to the equation $10^x = 500$.

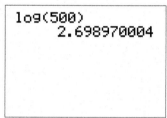

Figure 4. Calculating a common logarithm.

If the amount of energy released from one earthquake was 500 times greater than the amount of energy released from another, then, according to the rules of the Richter scale, the difference in magnitudes was about 2.699.

▶ Example 5

Finding the Value of a Common Logarithm Using a Calculator

1. Consider $\log(3215)$. Name the two integers that are above and below the value of the logarithm. Then find $\log(3215)$ to four decimal places using a calculator.

2. Consider $\log(7)$. Name the two integers that are above and below the value of the logarithm. Then find $\log(7)$ to four decimal places using a calculator.

Practice Set A — Answers

1. $\log_3(27) = 3$

2. $\log_6(36) = 2$

3. $\log_2(32) = 5$

4. $\log_4\left(\frac{1}{16}\right) = -2$

5. $\log_5\left(\frac{1}{125}\right) = -3$

6. $\log_7(\sqrt{7}) = \frac{1}{2}$

Solution

1. Note that $10^3 = 1000$ and $10^4 = 10,000$. Since 3215 is between 1000 and 10,000, we know that log(3215) is between log(1000) and log(10000). In other words, log(3215) is between 3 and 4.

1000	<	3215	<	10,000
3	<	log(3215)	<	4

Using a calculator, we find that log(3215) ≈ 3.5072.

2. Note that $10^0 = 1$ and $10^1 = 10$. Since 7 is between 1 and 10, we know that log(7) must be between log(1) and log(10). That is log(7) is between 0 and 1.

1	<	7	<	10
0	<	log(7)	<	1

Using a calculator we find log(7) ≈ 0.8451.

Practice Set B

Now it's your turn to work with a common logarithm. When you're finished, turn the page to check your work.

7. Without using a calculator, evaluate $y = \log(1,000,000)$.

8. Using a calculator, evaluate $y = \log(123)$ to four decimal places.

9. The amount of energy released from one earthquake was 8,500 times greater than the amount of energy released from another. The equation $10^x = 8500$ represents this situation, where x is the difference in magnitudes on the Richter scale. To the nearest thousandth, what was the difference in magnitudes?

C. BASIC PROPERTIES OF LOGARITHMS

Let's look at some basic properties of logarithms. We'll assume that $b > 0$ and $b \neq 1$. Now consider $\log_5(5)$ and $\log_{12}(12)$. The value of both is 1 because $5^1 = 5$ and $12^1 = 12$. In general, we know that $b^1 = b$, so $\log_b(b) = 1$.

Next, consider $\log_5(1)$ and $\log_{12}(1)$. The value of both logarithms is 0 because $5^0 = 1$ and $12^0 = 1$. In general, we know $b^0 = 1$, so $\log_b(1) = 0$.

We've seen that a logarithm is actually a function or rule that returns an exponent. For example, $\log_2(x)$ is a function that pairs the input, x, with an output that is the exponent on base 2 that equals x. We write the general definition of a logarithmic function as you see in the highlighted definition.

> **Properties of Logarithms**
>
> For $b > 0$ and $b \neq 1$,
> ♦ $\log_b(b) = 1$
> ♦ $\log_b(1) = 0$

The base of an exponential function is always positive, so no power of that base can ever be negative. The log of a negative number is not a real number. In addition, we cannot take the logarithm of zero. In general, the domain of a logarithmic function $\log_b(x)$ is the set of all positive real numbers or $(0, \infty)$.

> **Logarithmic Function**
>
> For $x > 0$, $b > 0$, $b \neq 1$,
> $y = \log_b(x)$ if and only if $x = b^y$.
>
> The function given by $f(x) = \log_b(x)$ is called the **logarithmic function with base b**.

▶ *Example 6*

Evaluating a Logarithmic Function

Let $f(x) = \log_3(x)$.

1. Find $f(9)$.   3. Find $f(3)$.
2. Find $f(81)$.   4. Find $f(1)$.

Solution

1. $f(9) = \log_3(9) = 2$ because $3^2 = 9$

2. $f(81) = \log_3(81) = 4$ because $3^4 = 81$

3. $f(3) = \log_3(3) = 1$ because $\log_b(b) = 1$

4. $f(1) = \log_3(1) = 0$ because $\log_b(1) = 0$

Notice we justified the last two solutions in Example 6 with the two basic properties of logarithms introduce in this section.

Let's look at two more properties of logarithms that emphasize the special relationship between exponential and logarithmic functions. This relationship is similar to the relationship between addition and subtraction or between multiplication and division. For example, if we have a quantity x and first add 7 to it and then subtract 7 from it, we get back x. Subtraction undoes addition:

$$x + 7 - 7 = x$$

If we have a quantity x and first multiple it by 5 and then divide it by 5, we get back x. Division undoes multiplication. Similarly, if we have a quantity x and first raise base 2 to x and then take the log base 2 of that expression, we get back x. Taking the logarithm undoes raising the base to the exponent.

$$\log_2(2^x) = x$$

> **Inverse Properties of Exponential and Logarithmic Functions with the Same Base**
>
> For $b > 0$ and $b \neq 1$,
> - $\log_b(b^x) = x$
> - $b^{\log_b(x)} = x$, where $x > 0$

Furthermore, if we have a quantity x and take the log base 9 of x, then raise 9 to that expression, we get back x. Raising the base to the exponent undoes taking the logarithm.

$$9^{\log_9(x)} = x$$

These two actions — taking the log base b and raising an exponent on base b — are inverse operations. This way that logarithms and exponential terms undo one another only works if they have the same base.

Practice Set B — Answers

7. $\log(1{,}000{,}000) = 6$

8. $\log(123) \approx 2.0899$

9. The equation $10^x = 8500$ is equivalent to $\log(8500) = x$. We find $\log(8500) \approx 3.929$. The difference in magnitudes was about 3.929 on the Richter scale.

▶ *Example 7*

Using the Inverse Properties of Exponential and Logarithmic Functions

Simplify.

1. $\log_6(6^9)$

2. $5^{\log_5(22)}$

3. $\log(10^{4.3})$

Solution

1. $\log_6(6^9) = 9$ because $\log_b(b^x) = x$.

2. $5^{\log_5(22)} = 22$ because $b^{\log_b(x)} = x$

3. This is a common logarithm, so the base is 10.
   $\log(10^{4.3}) = 4.3$ because $\log_b(b^x) = x$.

Practice Set C

Now it's your turn to work with logarithms. When you're finished, turn the page to check your work.

Let $f(x) = \log_2(x)$.

10. Find $f(16)$.   **11.** Find $f(\frac{1}{4})$.   **12.** Find $f(1)$

Simplify.

13. $8^{\log_8(512)}$ **14.** $\log(10^{1.33})$ **15.** $\log_{11}(11^4)$

Exercises

Find the logarithm.

1. $\log_5(125)$

2. $\log_4(16)$

3. $\log_3(243)$

4. $\log_6(216)$

5. $\log_8(64)$

6. $\log_2(128)$

7. $\log_4(256)$

8. $\log_5(625)$

9. $\log(10,000,000)$

10. $\log(100)$

11. $\log(1)$

12. $\log(10)$

13. $\log_3\left(\frac{1}{27}\right)$

14. $\log_2\left(\frac{1}{32}\right)$

15. $\log_6\left(\frac{1}{36}\right)$

16. $\log_8\left(\frac{1}{8}\right)$

17. $\log\left(\frac{1}{100,000}\right)$

18. $\log\left(\frac{1}{100}\right)$

19. $\log_{13}(1)$

20. $\log_6(1)$

21. $\log_7(7)$

22. $\log_3(3)$

23. $\log_{16}(4)$

24. $\log_{49}(7)$

25. $\log_{125}(5)$

26. $\log_{64}(4)$

27. $\log_6(\sqrt{6})$

28. $\log_3(\sqrt{3})$

29. $\log_8(\sqrt[4]{8})$

30. $\log_{16}(\sqrt[5]{16})$

31. $\log_{11}(11^5)$

32. $\log_9(9^7)$

For the following exercises, evaluate each expression using a calculator. Round to the nearest thousandth.

33. $\log(6,775)$

34. $\log(89,900)$

35. $\log(29)$

36. $\log(333)$

37. $\log(.249)$

38. $\log(0.058)$

Let $g(x) = \log_4(x)$

39. Find $g(256)$.

40. Find $g(64)$.

41. Find $g(\frac{1}{16})$.

42. Find $g(2)$.

Simplify.

43. $7^{\log_7(30)}$

44. $16^{\log_{16}(100)}$

45. $\log_8(8^5)$

46. $\log_2(2^7)$

47. $\log(10^6)$

48. $\log(10^3)$

49. $10^{\log(4)}$

50. $10^{\log(9)}$

Practice Set C — Answers

10. $f(16) = \log_2(16) = 4$

11. $f(\frac{1}{4}) = \log_2(\frac{1}{4}) = -2$

12. $f(1) = \log_2(1) = 0$

13. $8^{\log_8(512)} = 512$

14. $\log(10^{1.33}) = 1.33$

15. $\log_{11}(11^4) = 4$

3.2 Properties of Logarithms

OVERVIEW

In this section we will study some more properties of logarithms. We'll focus on how to use these properties to solve exponential and logarithmic equations. As you study this section, you will learn how to:

- Convert between exponential and logarithmic form.
- Use the power rule for logarithms.
- Use properties of logarithms to solve logarithmic equations.
- Use properties of logarithms to solve exponential equations.
- Use the change-of-base formula for logarithms.

A. CONVERTING BETWEEN EXPONENTIAL AND LOGARITHMIC FORMS

In Section 3.1, we learned how to find logarithms, and we often justified our solutions using the exponential form of the equation. For example, to find $\log_3 81$, we said:

$$\log_3 81 \ = \ 4 \text{ because } 3^4 \ = \ 81$$

The equation $\log_3 81 \ = \ 4$ is in **logarithmic form**, and the equation $3^4 \ = \ 81$ is in **exponential form**. Most importantly, these two forms are equivalent, and either one can replace the other when solving a problem.

Figure 1 offers more examples of equivalent equations in both logarithmic and exponential form.

> ### Logarithmic and Exponential Forms Property
> For $a > 0$, $b > 0$, $b \neq 1$, the equations $\log_b (a) = c$ and $b^c = a$ are equivalent.

Logarithmic Form	Exponential Form
$\log_b (a) = c$	$b^c \ = \ a$
$\log_4 (16) \ = \ 2$	$4^2 \ = \ 16$
$\log_5 (125) \ = \ 3$	$5^3 \ = \ 125$
$\log_2 (64) \ = \ 6$	$2^6 \ = \ 64$
$\log(100,000) \ = \ 5$	$10^5 \ = \ 100,000$

Figure 1.

▶ *Example 1*

Converting from Logarithmic Form to Exponential Form

Write the following logarithmic equations in exponential form.

1. $\log_6(216) = 3$ 3. $\log\left(\frac{1}{100}\right) = -2$

2. $\log_3(243) = 5$ 4. $\log_q(k) = m$

Solution

1. $\log_6(216) = 3$ becomes $6^3 = 216$

2. $\log_3(243) = 5$ becomes $3^5 = 243$

3. $\log\left(\frac{1}{100}\right) = -2$ becomes $10^{-2} = \frac{1}{100}$ (Remember: common log has base 10.)

4. $\log_q(k) = m$ becomes $q^m = k$

▶ *Example 2*

Converting from Exponential Form to Logarithmic Form

Write the following exponential equations in logarithmic form.

1. $4^6 = 4096$ 3. $10^4 = 10,000$

2. $2^{3.808} \approx 14$ 4. $81^{1/2} = 9$

Solutions

1. $4^6 = 4096$ becomes $\log_4(4096) = 6$

2. $2^{3.808} \approx 14$ becomes $\log_2(14) \approx 3.808$

3. $10^4 = 10,000$ becomes $\log(10,000) = 4$

4. $81^{1/2} = 9$ becomes $\log_{81}(9) = \frac{1}{2}$

Practice Set A

Write the following logarithmic equations in exponential form.

1. $\log_{10}(1,000,000) = 6$ 2. $\log_5(25) = 2$

Write the following exponential equations in logarithmic form.

3. $3^2 = 9$ 4. $5^3 = 125$ 5. $2^{-1} = \frac{1}{2}$

Turn the page to check your solutions.

B. SOLVING EQUATIONS IN LOGARITHMIC FORM

Converting between equations in exponential and logarithmic forms is an essential skill for solving equations involving exponents or logarithms. The conversion often turns a seemingly impossible problem into a very simple problem to solve.

▶ *Example 3*

Solving Equations in Logarithmic Form

Solve for x.

1. $\log_6(x) = 4$

2. $\log_2(5x - 4) = 8$

Solution

1. We write $\log_6(x) = 4$ in exponential form and simplify:

$6^4 = x$	Write in exponential form.
$1296 = x$	Simplify.

2. We write $\log_2(5x - 4) = 8$ in exponential form and solve for x:

$2^8 = 5x - 4$	Write in exponential form.
$256 = 5x - 4$	Simplify.
$260 = 5x$	Add 4 to both sides.
$52 = x$	Divide both sides by 5.

Many of the solving steps we need to carry out to solve logarithmic equations are familiar to us from previous algebra courses. It's important to execute the solving steps in the correct order. When we are solving, we're "undoing" the equation to get at x. Therefore, the order of the steps is the reverse of PEMDAS, the correct order of operations for computations. In shorthand, the correct order of operations for solving is SADMEP, as you see in Figure 2.

We group addition with subtraction, we group multiplication with division, and likewise, we group logarithms with exponents

Order of Operations for Computations (PEMDAS)		Order of Operations for Solving Equations (SADMEP)	
P	expressions in Parentheses	S	Subtraction (and addition)
E	Exponents (and logarithms)	A	Addition (and subtraction)
M	Multiplication (and division from left to right)	D	Division (and multiplication)
D	Division (and multiplication from left to right)	M	Multiplication (and division)
A	Addition (and subtraction from left to right)	E	Exponents (and logarithms)
S	Subtraction (and addition from left to right)	P	expressions in Parentheses

Figure 2. The correct order of operations.

when following the correct order of operations. This is true both for computations and for solving equations. Pay attention to the order of the steps as we work the next examples.

▶ *Example 4*

Solving Logarithmic Equations

Solve for x.

   1. $\log_3(2x-5)+7 = 11$

   2. $4\log(x-7)+1 = 9$

Solution

1. We get $\log_3(2x-5)$ alone on the left side of the equation before we write it in exponential form and solve for x:

$\log_3(2x-5)+7 = 11$	Original equation.
$\log_3(2x-5) = 4$	Subtract 7 from both sides.
$3^4 = 2x-5$	Write in exponential form.
$81 = 2x-5$	Simplify.
$86 = 2x$	Add 5 to both sides.
$43 = x$	Divide both sides by 2.

2. We get $\log(x-7)$ alone on the left side of the equation before we write it in exponential form and solve for x:

$4\log(x-7)+1 = 9$	Original equation.
$4\log(x-7) = 8$	Subtract 1 from both sides.
$\log(x-7) = 2$	Divide both sides by 4.
$10^2 = x-7$	Write in exponential form. For a common logarithm the base is 10.
$100 = x-7$	Simplify.
$107 = x$	Add 7 to both sides.

In the next example, we'll solve a logarithmic equation for the base. Again, we write the equation in exponential form as part of the solving process. Then we use a reciprocal exponent to solve for b.

Practice Set A — Answers

1. $10^6 = 1,000,000$ **3.** $\log_3(9) = 2$ **5.** $\log_2(\frac{1}{2}) = -1$

2. $5^2 = 25$ **4.** $\log_5(125) = 3$

▶ *Example 5*

Solving for the Base of a Logarithm

Solve for b.

1. $\log_b(243) = 5$ 2. $\log_b(59) = 4$

Solution

1. We write $\log_b(243) = 5$ in exponential form and solve for b:

   $b^5 = 243$ Write in exponential form.

   $b = 243^{1/5}$ The solution of $b^5 = k$ is $\pm k^{1/5}$.

   $b = 3$ Simplify.

2. We write $\log_b(59) = 4$ in exponential form and solve for b:

   $b^4 = 59$ Write in exponential form.

   $b = 59^{1/4}$ The solution of $b^4 = k$ is $\pm k^{1/4}$.

   $b \approx \pm 2.7715$ Simplify. The exponent was even, so there are two solutions.

   $b \approx 2.7715$ The base of a logarithm is positive.

Practice Set B

Solve for x. When you're finished, turn the page to check your answers.

6. $\log_2(x) = 9$ 7. $\log_3(2x+1) = 5$ 8. $7\log(x) - 24 = 18$

Solve for b.

9. $\log_b(729) = 3$ 10. $\log_b(311) = 6$

C. USING THE POWER RULE FOR LOGARITHMS TO SOLVE EXPONENTIAL EQUATIONS

Our objective is to be able to solve exponential equations with the variable in the exponent. Here are some examples of exponential equations:

$5^x = 94$

$2(1.048)^x = 6.6$

$3^{2x+1} = 774$

The Power Rule for Logarithms
For $a > 0$, $b > 0$, and $b \neq 1$, $\log_b(a^p) = p\log_b(a)$

The **power rule for logarithms** states that a logarithm of a power is equal to the exponent times the logarithm. This property is useful for solving exponential equations. We will first investigate the power rule of logarithms, and then we'll use it to help us solve exponential equations for the variable. It's easier to understand in formula form, as you see in the highlighted box, and it's a simple process when we apply it.

▶ *Example 6*

Applying the Power Rule to a Logarithm

Apply the power rule of logarithms to rewrite the following expressions.

    1. $\log_2(x^5)$ **2.** $\log_6(3.4^t)$ **3.** $\log(3^{x-1})$

Solution

1. $\log_2(x^5) = 5\log_2(x)$

2. $\log_6(3.4^t) = t\log_6(3.4)$

3. $\log(3^{x-1}) = (x-1)\log(3)$

Some math textbooks include a proof of this important property. We haven't done that, but we will point out that the power rule for logarithms follows the rule for exponents regarding raising a power to a power. This property for exponents states that "when raising an exponent to a power, multiply the exponents":

$$(b^m)^n = b^{mn}$$

When raising a logarithm to an exponent, we also rewrite the expression as a multiplication. After all, a logarithm is an exponent.

Next, let's look at the **logarithm property of equality**, which is also useful in solving exponential equations.

This property states that if we have an equation, such as $a = c$, we can take the log base b of both sides of the equation. This property is true for *any* base b. In practice, however, we will use base 10, the common logarithm, because our calculators are programmed to find the logarithm of any positive number in base 10. When solving exponential equations, we refer to this step simply as "taking the log of both sides."

Logarithm Property of Equality
For positive real numbers, a, b, and c, where $b \neq 1$, the following equations are equivalent. $a = c$ and $\log_b(a) = \log_b(c)$

Now let's use these two new properties of logarithms to solve some equations.

▶ *Example 7*

Solving an Exponential Equation

Solve the equation $5^x = 402$.

Practice Set B — Answers

6. $2^9 = x \rightarrow x = 512$

7. $3^5 = 2x + 1 \rightarrow x = 121$

8. $\log(x) = 6 \rightarrow 10^6 = x \rightarrow x = 1{,}000{,}000$

9. $b^3 = 729 \rightarrow b = 9$

10. $b^6 = 311 \rightarrow b \approx 2.6029$

Solution

$$5^x = 402 \qquad \text{Original equation.}$$

$$\log(5^x) = \log(402) \qquad \text{Take the log of both sides.}$$

$$x\log(5) = \log(402) \qquad \text{Power rule for logarithms.}$$

$$x = \frac{\log(402)}{\log(5)} \qquad \text{Divide both sides by } \log(5). \text{ Make sure to use both sets of parentheses on the calculator!}$$

$$x \approx 3.7258 \qquad \text{Compute.}$$

We can check that 3.7258 is the approximate solution to the equation $5^x = 402$:

$$5^{3.7258} \approx 401.9967 \approx 402$$

▶ *Example 8*

Solving an Exponential Equation

Solve the equation $2(3)^x = 52$.

Solution

$$2(3)^x = 52 \qquad \text{Original equation.}$$

$$(3)^x = 26 \qquad \text{Divide both sides by 2.}$$

$$\log(3^x) = \log(26) \qquad \text{Take the log of both sides.}$$

$$x\log(3) = \log(26) \qquad \text{Apply the power rule for logarithms.}$$

$$x = \frac{\log(26)}{\log(3)} \qquad \text{Divide both sides by } \log(3).$$

$$x \approx 2.9656 \qquad \text{Compute.}$$

We can check that 2.9656 is the approximate solution to the equation $2(3)^x = 52$:

$$2(3)^{2.9656} \approx 51.9973 \approx 52$$

▶ *Example 9*

Solving an Exponential Equation

Solve the equation $2^{4x-1} = 294$.

Solution

$$2^{4x-1} = 294 \qquad \text{Original equation.}$$

$$\log(2^{4x-1}) = \log(294) \qquad \text{Take the log of both sides.}$$

$$(4x-1)\log(2) = \log(294) \qquad \text{Apply the power rule for logarithms. The expression on the left is not correct without the parentheses.}$$

$$4x - 1 = \frac{\log(294)}{\log(2)} \qquad \text{Divide both sides by } \log(2).$$

$$4x = \frac{\log(294)}{\log(2)} + 1 \qquad \text{Add 1 to both sides.}$$

$$x = \frac{\frac{\log(294)}{\log(2)} + 1}{4} \qquad \text{Divide both sides by 4.}$$

$$x \approx 2.2999 \qquad \text{Compute.}$$

We can check that 2.9656 is the approximate solution to the equation $2^{4x-1} = 294$:

$$2^{4(2.2999)-1} \approx 293.9853 \approx 294$$

After the fourth line in the solving process above, it gets kind of messy to continue solving without calculating the division of the logarithms. We could round the expression $\frac{\log(294)}{\log(2)}$ at this point in the process, but we may lose some accuracy in our final answer.

We can enter the calculations on our calculator one step at a time, as you see in Figure 3, or we can enter it all at once at the last step, as you see in Figure 4. Always be careful to use parentheses correctly.

```
log(294)/log(2)
          8.199672345
Ans+1
          9.199672345
Ans/4
          2.299918086
█
```

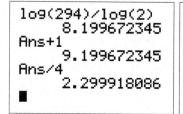

Figure 3. Computing one step at a time.

```
(log(294)/log(2)
+1)/4
          2.299918086
█
```

Figure 4. Computing at the last step.

Example 10

Solving an Exponential Equation

Solve the equation $11(.25)^x - 8 = 47$.

Solution

$11(1.25)^x - 8 = 47$		Original equation.
$11(1.25)^x = 55$		Add 8 to both sides.
$(1.25)^x = 5$		Divide both sides by 11.
$\log(1.25)^x = \log(5)$		Take the log of both sides.
$x\log(1.25) = \log(5)$		Apply the power rule for logarithms.
$x = \frac{\log(5)}{\log(1.25)}$		Divide both sides by log(1.25).
$x \approx 7.2126$		Compute.

We can check that 7.2126 is the approximate solution to $11(1.25)^x - 8 = 47$:

$$11(1.25)^{7.2126} - 8 \approx 47.0004 \approx 47$$

Practice Set C

Solve for x. Round answers to the fourth decimal place. Turn the page to check your solutions.

11. $4^x = 555$

12. $3(9)^x = 6810$

13. $5^{3x-2} = 147$

14. $822 = 7(2.5)^x - 4$

D. SOLVING EQUATIONS BY USING GRAPHS

We can also solve exponential equations graphically. In Example 10, we solved the following equation algebraically: $11(1.25)^x - 8 = 47$. The solution was $x \approx 7.2126$.

Now let's solve this equation graphically. Use the system of equations below, where y_1 equals the left side of the equation and y_2 equals the right side of the equation. After finding a good window, use the intersect feature on the calculator to solve for x. Figure 5 shows you how it's done.

$$y_1 = 11(1.25)^x - 8$$
$$y_2 = 47$$

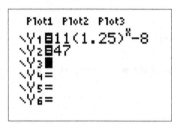
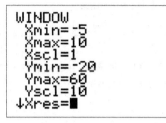
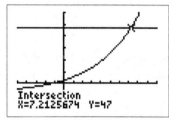

Figure 5.

Notice we find the same approximate solution from graphing, $x \approx 7.2126$, that we found by solving algebraically. Some equations containing exponential terms are impossible to solve algebraically but can be solved graphically. The equation in the next example is of this type.

▶ *Example 11*

Solving an Equation with an Exponential Term Graphically

Solve $2^{x-4} + 3 = 4x - 1$.

Solution

The equation has an exponential term and a linear term. It is impossible to solve algebraically. To solve graphically, enter the system of equations below and use the intersect feature to solve. See Figure 6.

$$y_1 = 2^{x-4} + 3$$
$$y_2 = 4x - 1$$

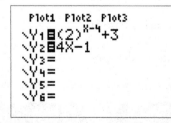
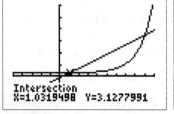
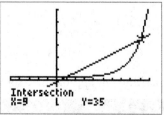

Figure 6. Solve the equation by graphing.

There are two solutions because the curves intersect at two points. The points have (approximate) coordinates (1.0319, 3.1278) and (9, 35). These x-y pairs are solutions to the system of equations. The solutions to the original equation are $x \approx 1.0319$ and $x = 9$.

E. CHANGE-OF-BASE

In order to evaluate logarithms with a base other than 10, we can use the **change-of-base formula**. This formula allows us to rewrite the logarithm as the quotient of logarithms of any other base. When we're using a calculator to solve an exponential equation, we change the base to common log because our calculators are programmed to evaluate the log base 10 of any positive number.

> ### Change-of-Base Property
>
> For positive real numbers a, b, and c, where $b \neq 1$ and $c \neq 1$,
> $$\log_b(a) = \frac{\log_c(a)}{\log_c(b)}$$

Here are some examples of the change of base formula:

$$\log_2(14) = \frac{\log_3(14)}{\log_3(2)} \qquad \log_4(x) = \frac{\log_9(x)}{\log_9(4)} \qquad \log_5(36) = \frac{\log(36)}{\log(5)}$$

We'll use this last example to take a closer look because the original logarithm was changed to a logarithm base 10, which we can compute on a calculator.

▶ *Example 12*

Converting to Log Base 10

Evaluate $\log_5 36$ using a calculator.

Solution

$$\log_5 36 = \frac{\log(36)}{\log(5)} \qquad \text{Apply the change of base formula using base 10.}$$

$$\approx 2.2266 \qquad \text{Use a calculator to evaluate to 4 decimal places.}$$

Checking our approximate solution with the exponential form we find that $5^{2.2266} \approx 36$.

▶ *Example 13*

Converting to Log Base 10

Evaluate $\log_{1.75} 86.38$ using a calculator.

Solution

$$\log_{1.75} 86.38 = \frac{\log(86.38)}{\log(1.75)} \qquad \text{Apply the change of base formula using base 10.}$$

$$\approx 7.9675 \qquad \text{Use a calculator to evaluate to 4 decimal places.}$$

Checking our approximate solution with the exponential form we find that $1.75^{7.9675} \approx 86.38$.

Practice Set C — Answers

11. $x = \dfrac{\log(555)}{\log(4)} \approx 4.5582$

12. $x = \dfrac{\log(2270)}{\log(9)} \approx 3.5170$

13. $x = \dfrac{\frac{\log(147)}{\log(5)} + 2}{3} \approx 1.7002$

14. $x = \dfrac{\log(118)}{\log(2.5)} \approx 5.2065$

To see how we can use the change-of-base formula to solve exponential equations, let's revisit a problem that we worked earlier in this section in Example 7. We will compare the original solving method with the method that uses the change-of-base formula.

▶ *Example 14*

Converting to Log Base 10

Solve the equation $5^x = 402$.

Solution

Method 1: Take the log and both sides and then use the power rule for logarithms.

$5^x = 402$	Original equation.
$\log(5^x) = \log(402)$	Take the log of both sides.
$x \log(5) = \log(402)$	Apply the power rule for logarithms.
$x = \dfrac{\log(402)}{\log(5)}$	Divide both sides by $\log(5)$.
$x \approx 3.7258$	Compute.

Method 2: Convert to logarithmic form and then use the change-of-base formula.

$5^x = 402$	Original equation.
$\log_5(402) = x$	Write the equation in logarithmic form.
$\dfrac{\log(402)}{\log(5)} = x$	Change-of-base formula: convert to base 10.
$3.7258 \approx x$	Compute.

We arrive at the same solution with either method.

▶ *Example 15*

Using the Change-of-Base Property to Solve an Exponential Equation

Solve for x: $4(7)^x = 58$

Solution

$4(7)^x = 58$	Original equation.
$(7)^x = 14.5$	Divide both sides by 4.
$\log_7(14.5) = x$	Write the equation in logarithmic form.
$\dfrac{\log(14.5)}{\log(7)} = x$	Apply the change-of-base formula.
$1.2345 \approx x$	Compute.

F. EXPONENTIAL MODELS

In 1859, an Australian landowner named Thomas Austin released 24 rabbits into the wild for hunting. Because Australia had few predators and ample food, the rabbit population exploded. In fewer than ten years, the rabbit population numbered in the millions.

Uncontrolled population growth, as in the wild rabbits in Australia, can be modeled with exponential functions. Equations resulting from those exponential functions can be solved to analyze and make predictions about exponential growth. To wrap up this section, we will now use these techniques for solving exponential functions and apply them to real-world situations.

Recall from Section 2.5 that for models of the form $f(t) = ab^t$, the initial value is a and $b = 1 \pm r$ where r is the percent change in decimal form. This model has many real-world applications including problems involving investments with compounded interest.

▶ *Example 16*

Predicting the Value of an Investment Compounded Annually

Let $B = f(t)$ be the balance of a bank account after t years with an initial investment of $5000 and an annual interest rate of 2.4%.

   a. Write an exponential equation to model this situation.

   b. Find and interpret $f(7)$.

   c. To the nearest tenth of a year, predict when the balance of the account will be $8,000.

Solution

a. The initial value is $a = 5000$. The base is $b = 1 + 0.024 = 1.024$: $B = f(t) = 5000(1.024)^t$

b. $f(7) = 5000(1.024)^7 \approx 5902.96$. After 7 years, the balance in the account will be $5902.96.

c. Replace $f(t)$ with $8000 and solve for x.

$5000(1.024)^t = 8000$	$f(t) = 8000$	
$1.024^t = 1.6$	Divide both sides by 5000.	
$\log(1.024^t) = \log(1.6)$	Take the log of both sides.	
$t\log(1.024) = \log(1.6)$	Apply the power rule for logarithms.	
$t = \dfrac{\log(1.6)}{\log(1.024)}$	Divide both sides by $\log(1.024)$.	
$t \approx 19.8$	Compute.	

It will take approximately 19.8 years for the balance of the account to grow to $8000.

In Section 2.5, Example 7, we found an exponential regression equation to model the world population in billions of people. We use this model again for the example below.

▶ Example 17

Predicting Population Growth

Let $f(t)$ represent the world's population in billions t years after 1900:

$$f(t) = 1.21(1.0161)^t$$

Estimates vary, but many experts think the carrying capacity of the planet is about 10 billion people (unless we all stop eating meat). Predict when the world population will grow to 10 billion.

Solution

Replace $f(t)$ with 10 and solve for t.

$1.21(1.0161)^t = 10$	$f(t) = 10$
$1.0161^t = \dfrac{10}{1.21}$	Divide both sides by 1.21.
$\log(1.0161^t) = \log\left(\dfrac{10}{1.21}\right)$	Take the log of both sides.
$t\log(1.0161) = \log\left(\dfrac{10}{1.21}\right)$	Apply the power rule for logarithms.
$t = \dfrac{\log\left(\dfrac{10}{1.21}\right)}{\log(1.0161)}$	Divide both sides by $\log(1.0161)$.
$t \approx 132.23$	Compute.

Approximately 132 years after 1900, the world population will grow to 10 billion.

$$1900 + 132 = 2032$$

We estimate the planet will reach its carrying capacity in the year 2032.

▶ Example 18

Predicting a Depreciation Value

A 2016 Tesla Model X costs approximately $95,000 when new. It's expected that this unique car will depreciate less quickly than the average automobile. However, the Model X still depreciates in value by about 8.5% per year.

a. Write an equation that models value of a Tesla Model X t years after it is purchased.

b. What is the value of the car after 3 years?

c. Predict to the nearest tenth of a year when the Model X will be worth only $30,000.

Solution

a. Let $f(t)$ represent the value of the Tesla Model X t years after it is purchased. The initial value is $a = 95{,}000$. The value of the base is $b = 1 - 0.085 = 0.915$:

$$f(t) = 95{,}000(0.915)^t$$

b. $f(3) = 95,000(0.915)^3 \approx 72,776$

After 3 years the Tesla Model X will be worth about $72,776.

c. Replace $f(t)$ with $30,000 and solve for t.

$$95,000(0.915)^t = 30,000 \qquad f(t) = 30,000$$

$$0.915^t = \frac{6}{19} \qquad \text{Divide both sides by 95,000.}$$

$$\log(0.915^t) = \log\left(\frac{6}{19}\right) \qquad \text{Take the log of both sides.}$$

$$t\log(0.915) = \log\left(\frac{6}{19}\right) \qquad \text{Apply the power rule for logarithms.}$$

$$t = \frac{\log\left(\frac{6}{19}\right)}{\log(0.915)} \qquad \text{Divide both sides by } \log(0.915).$$

$$t \approx 12.976 \qquad \text{Compute.}$$

In approximately 13 years, the value of the Tesla Model X will depreciate to $30,000.

Note: if you round the value on the second step, make sure to hold at least 3 decimal places for accuracy in the final answer.

▶ *Example 19*

Finding Doubling Time

When twins Julio and Rodrigo turned 21, they received an inheritance of $10,000 each from their grandfather. Julio spent most of his inheritance right away, but before it was all gone, he set aside $500 to invest in a mutual fund. Rodrigo invested all of his inheritance in the same mutual fund which had a 6.75% interest rate compounded annually.

    a.   Write an exponential equation to model Julio's investment.

    b.   Determine the time it takes for Julio's initial investment to double.

    c.   Compare the answer to (b) to the amount of time it takes Rodrigo's investment to double.

Solution

a. Let $f(t)$ represent the value of the investment after t years. Julio's initial investment is $500. The value of the base $b = 1 + 0.0675 = 1.0675$:

$$f(t) = 500(1.0675)^t$$

b. When $500 is doubled it equals $1000. Replace $f(t)$ with 1000 and solve for t.

$$500(1.0675)^t = 1000 \qquad f(t) = 1000$$

$$1.0675^t = 2 \qquad \text{Divide both sides by 500.}$$

$$\log(1.0675^t) = \log(2) \qquad \text{Take the log of both sides.}$$

$$t\log(1.0675) = \log(2) \qquad \text{Apply the power rule for logarithms.}$$

$$t = \frac{\log(2)}{\log(1.0675)} \qquad \text{Divide both sides by } \log(1.0675).$$

$$t \approx 10.6 \qquad \text{Compute.}$$

It will take about 10.6 years for Julio's investment of $500 to double.

c. Rodrigo's investment can be modeled by $f(t) = 10,000(1.0675)^t$.

Since the brothers invested in the same fund, the only part of the equations modeling the growth of their investments that has changed is the initial value.

To find the doubling time for Rodrigo's $10,000 investment, replace $f(t)$ with 20,000 and solve for t.

$$10,000(1.0675)^t = 20,000 \qquad\qquad f(t) = 20,000$$

$$1.0675^t = 2 \qquad\qquad \text{Divide both sides by 10,000.}$$

$$\log(1.0675^t) = \log(2) \qquad\qquad \text{Take the log of both sides.}$$

$$t\log(1.0675) = \log(2) \qquad\qquad \text{Apply the power rule for logarithms.}$$

$$t = \frac{\log(2)}{\log(1.0675)} \qquad\qquad \text{Divide both sides by} \log(1.0675).$$

$$t \approx 10.6 \qquad\qquad \text{Compute.}$$

After the first step, the calculation is exactly the same as it was in Julio's case! It takes about 10.6 years for Rodrigo's $10,000 investment to double.

What we see in this example is something interesting about the behavior of exponential growth functions. The doubling time is the same no matter where you begin in the growth process.

Exercises

Write the following logarithmic equations in exponential form. Assume all constants are positive and not equal to 1.

1. $\log_4(64) = 3$
2. $\log_{11}(121) = 2$
3. $\log_5(625) = 4$
4. $\log_3(243) = 5$

5. $\log_2\left(\frac{1}{8}\right) = -3$
6. $\log_9\left(\frac{1}{81}\right) = -2$
7. $\log(1) = 0$
8. $\log(1000) = 3$

9. $\log_w(m) = r$
10. $\log_z(y) = k$

Write the following exponential equations in logarithmic form. Assume all constants are positive and not equal to 1.

11. $2^8 = 256$
12. $5^6 = 15625$
13. $49^{1/2} = 7$

14. $27^{1/3} = 3$
15. $10^4 = 10,000$
16. $10^{-1} = 0.1$

17. $4^x = 97$
18. $12^x = 44$

Solve for x.

19. $\log_{12}(x) = 2$
20. $\log_7(x) = 3$

21. $\log_8(x) = 0$
22. $\log_{11}(x) = 0$

23. $\log(x) = -3$
24. $\log(x) = -2$

25. $\log_2(7x + 4) = 5$

26. $\log_3(10x - 9) = 4$

27. $6\log_5(x) - 15 = -3$

28. $5\log_4(x) + 11 = 26$

29. $24\log_4(x) - 17 = -5$

30. $3\log_8(x) + 17 = 18$

Solve for *b*. If needed, round solutions to the fourth decimal place.

31. $\log_b(2401) = 4$

32. $\log_b(216) = 3$

33. $\log_b(9) = \frac{1}{2}$

34. $\log_b(1000) = \frac{1}{3}$

35. $\log_b(188) = 7$

36. $\log_b(75) = 6$

Solve for *x*. If needed, round solutions to the fourth decimal place. Check your solutions in the original equation.

37. $3^x = 101$

38. $6^x = 340$

39. $4(5)^x = 708$

40. $56 = 2(9)^x$

41. $17.16 = 6.24(1.15)^x$

42. $9.25(1.065)^x = 31.82$

43. $7(4)^x - 55 = 241.8$

44. $98 = 3(5)^x + 40$

45. $3^{2x+4} = 695$

46. $4^{3x-7} = 879$

47. $4000 = 25(1.2)^{3x}$

48. $38(1.04)^{x+1} = 133$

49. $50(2)^{x-3} + 66 = 786$

50. $7.7(3)^{2x} - 48 = 337$

Use the "intersect" feature on a graphing calculator to solve the following equations. If needed, round solutions to the fourth decimal place.

51. $4^x + 3 = \frac{1}{5}x + 8$

52. $3^x - 5 = -2x + 7$

53. $1.5x - 2 = 2.4^{x-3}$

54. $-\frac{1}{2}x + 4 = 0.8^{2x-7}$

55. $0.49^x = -3x - 8$

56. $1.4^x = x - 6$

Apply what you have learned in this chapter to solve the following real-world applications.

57. An investment of $2200 is made in a stock that increases in value by about 4.7% per year.

   a. Write an equation to model the value of the investment over *t* years.

   b. Use the model to predict the value of the investment in 12 years.

   c. How long it will take for the initial investment to be worth $5000?

58. An investment of $750 is made in an account that earns a 3.9% interest rate compounded yearly.

   a. Write an equation to model the value of the investment over *t* years.

   b. Use the model to predict the value of the investment in 7 years.

   c. How long it will take for the initial investment to be worth $1200?

59. A person invests $14,000 in an account that earns 2.25% interest compounded annually. When will the investment be worth $25,000?

60. A person invests $8000 in an account that earns 7.45% interest compounded annually. When will the investment be worth $12,000?

61. The population of a species of whales is estimated to be decreasing at a rate of 4% per year. The current population is approximately 15,000.

   a. Write an equation to model the population of whales over *t* years.

   b. Use the model to predict the population after 10 years.

   c. When will the whale population be 8000?

62. The population of white owls in a wilderness area is estimated to be decreasing at a rate of 7% per year. The current population is approximately 98.

   a. Write an equation to model the population of owls over *t* years.

   b. Use the model to predict the population after 5 years.

   c. When will the owl population be below 50?

63. The value of a certain car model is $22,500 when new, but it depreciates at a rate of 16% per year. When will the value of the car depreciate to $10,000?

64. The value of a construction crane is $375,000 when new, but depreciates at a rate of 13% per year. When will the value of the crane depreciate to $150,000?

65. The following equation models the amount of caffeine in a person's bloodstream in milligrams, t hours after 7:00 a.m. $f(t) = 180\left(\frac{1}{2}\right)^{t/6}$

    a. How much caffeine is in the person's bloodstream at 7:00 a.m.?

    b. Find and interpret $f(12)$.

    c. When will the caffeine level be less than 10 milligrams?

66. The following equation models the amount of aspirin in a person's blood in milligrams, t hours after ingesting a dose: $f(t) = 300\left(\frac{1}{2}\right)^{t/20}$

    a. How much aspirin is in the dose?

    b. Find and interpret $f(60)$.

    c. When will the aspirin level be less than 10 milligrams?

67. The population of a small coastal town is modeled by the following equation where $f(t)$ represents the population in thousands t years after 2000: $f(t) = 7.8(1.013)^t$

    a. What does the value of the base mean in the context of the problem?

    b. Find and interpret $f(8)$.

    c. Predict when the population of the town will be 10 thousand.

68. A population of suburban town is model by the following equation where $f(t)$ represents the population in thousands t years after 2010: $f(t) = 33(1.021)^t$

    a. What does the value of the base mean in the context of the problem?

    b. Find and interpret $f(5)$.

    c. Predict when the population of the town will be 10 thousand.

69. The number of passengers traveling by airplane each year has increased according to the model, $f(t) = 466(1.035)^t$, where $f(t)$ represents the number of airline passengers in millions t years after 1990. *(US Census Bureau)*

    a. How many passengers travel by air in 2001?

    b. In what year is it predicted that 2 billion (2000 million) passengers will travel by airplane?

70. The average price of a ticket to a major league baseball game has increased according to the model, $f(t) = 1.22(1.051)^t$, where $f(t)$ represents the price of a ticket in dollars t years after 1950.

    a. What was the price of a ticket in 1960? In 2017?

    b. In what year is it predicted that the price of a ticket will be $50?

71. The number of bankruptcies (in millions) filed per year since 1994 can be modeled by the equation $f(t) = 0.798(1.164)^t$. *(Admin Office of the U.S. Courts)* In what year is it predicted that 50 million bankruptcies will be filed?

72. The amount of money (in billions of dollars) spent on health care in the U.S. per year since 1970 can be modeled by the equation $f(t) = 78.16(1.11)^t$. In what year is it estimated that $40 trillion will be spent on health care expenditures?

73. A venture capitalist invests $1.2 million into a start-up company whose sales pitch promises an 8% per year growth on the investment. How long will it take for the initial investment to double?

74. A start-up company is valued at $24.3 million and increases in value by 6% per year. How long will it take for the value of the company to double?

3.3 Natural Logarithms

OVERVIEW

In many of the sciences, the most frequently used base for logarithms is a special number, e. Base e logarithms, known as natural logarithms, and the related exponential expressions with base e are studied in this section. Equations using base e are uniquely suited to describing phenomenon in the natural world. The base e logarithm, $\log_e(x)$, has its own notation, $\ln(x)$.

As you study this section, you will learn to:

- ◆ Know the meaning of a natural logarithm.
- ◆ Evaluate natural logarithms.
- ◆ Use exponential models with base e to make estimates and predictions.

A. DEFINITION OF NATURAL LOGARITHM

A **natural logarithm** is a logarithm with a specific base called e, where e is an irrational number. Rounded to twenty decimal places:

$$e \approx 2.71828182845904523635 \ldots \approx 2.72$$

Notice the value of e is between the whole numbers 2 and 3. The letter e in this case does *not* represent a variable; rather, it represents a constant – the same number all the time. In this way it is similar to another irrational number that you may be familiar with, the number π (pi). The number π is represented by a Greek letter but is also a constant and *not* a variable.

We write $\log_e(x)$ simply as $\ln(x)$. We read $\ln(x)$ as "the logarithm with **base** e of x" or most commonly, "the natural log of x."

The working definition is that a natural logarithm, $y = \ln(x)$ is the exponent to which e must be raised to get x. The equation $y = \ln(x)$ is in logarithmic form and the equation $e^y = x$ is in exponential form.

For example, we can rewrite the logarithmic equation $y = \ln(99)$ in exponential form as $e^y = 99$.

Most values of $\ln(x)$ can only be found by using a calculator. The major exception, of course, is $\ln(1) = 0$ because the logarithm of 1 is always 0 in any base. For other natural logarithms, we can use the $\boxed{\text{LN}}$ key that can be found on your graphing calculator.

Given a natural logarithm with the form $y = \ln(x)$, evaluate it using a calculator by following these steps.

1. Press $\boxed{\text{LN}}$.

2. Enter the value given for x, followed by $\boxed{)}$.

3. Press $\boxed{\text{ENTER}}$.

> ### Natural Logarithm
>
> A **natural logarithm** is a logarithm with base e. We write $\ln(x)$ to represent $\log_e(x)$.
>
> For $x > 0$, $y = \ln(x)$ is equivalent to $e^y = x$.

To calculate a power of e, we use the [2nd] button to access the command e^x. Notice this command already sets up an exponent entry.

1. Press [2nd], then Press [LN].

2. Enter the value given for x, followed by).

3. Press [ENTER].

▶ Example 1

Finding a Natural Logarithm

Use a calculator to find $\ln(72)$.

Solution

We find that the calculator shows $\ln(72) \approx 4.2767$.

We use the e^x command to verify our solution. In this case,

$$\ln(72) \approx 4.2767, \text{ because } e^{4.2767} \approx 72$$

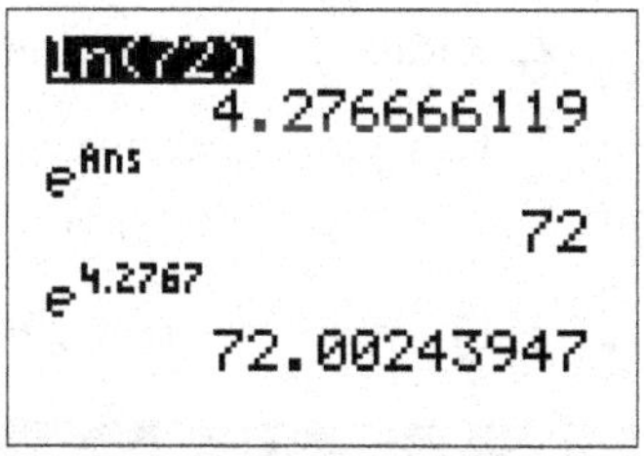

Figure 1 shows the calculation of $\ln(72)$ and its verification in exponential form. Notice that if we use the answer command, [2nd] [(-)], the exponential form returns exactly 72. If we round the value of the natural logarithm, the answer will be approximately 72.

Figure 1. Verifying ln(72).

The properties of logarithms base b that we studied in Sections 3.1 and 3.2 apply to natural logarithms as well. For positive numbers x and b, and $b \neq 1$, the following properties apply:

Properties of Logarithms	*Properties of Natural Logarithms*	
$\log_b(b) = 1$	$\ln(e) = 1$	The natural logarithm of e is 1.
$\log_b(1) = 0$	$\ln(1) = 0$	The natural logarithm of 1 is 0.
$\log_b(bx) = x$	$\ln(e^x) = x$	Inverse property of logarithms
$b^{\log_b(x)} = x$	$e^{\ln(x)} = x$	Inverse property of logarithms
$\log_b(a^p) = p\log_b(a)$	$\ln(x^p) = p\ln(x)$	Power rule of logarithms
If $a = c$, then $\log_b(a) = \log_b(c)$	If $a = c$, then $\ln(a) = \ln(c)$	Logarithm property of equality

Figure 2.

▶ *Example 2*

Finding a Natural Logarithm

Evaluate the expression without using a calculator.

    1. $\ln(e^7)$ **2.** $\ln(\sqrt{e})$ **3.** $e^{\ln(4)}$

Solution

1. Using an inverse property of logarithms we see

$$\ln(e^7) = \log_e(e^7) = 7.$$

In words, the expression $\ln(e^7)$ asks, "What exponent do we raise e to, so that it produces e^7?". The answer, of course, is 7.

2. Recall by the definition of fractional exponents that $\sqrt{x} = x^{1/2}$. So $\ln(\sqrt{e}) = \ln(e^{1/2}) = \frac{1}{2}$.

3. Using the other inverse property of logarithms we see,

$$e^{\ln(4)} = e^{\log_e(4)} = 4.$$

In words, "We raise e to the exponent that is the exponent on e that produces 4." The answer is 4.

Practice Set A

Use a calculator to find the logarithms to four decimal places. Write the solution in exponential form to verify your work.

    1. $\ln(400)$ **2.** $\ln(0.77)$ **3.** $\ln(-16)$

Evaluate the expression without using a calculator.

    4. $e^{\ln(38)}$ **5.** $\ln(\sqrt[3]{e})$ **6.** $\ln(e^2)$

Turn the page to check your answers.

B. SOLVING LOGARITHMIC AND EXPONENTIAL EQUATIONS IN BASE E

In Section 3.2 we solved equations in logarithmic form by rewriting the equation in exponential form. Remember, sometimes it is necessary to simplify the equation first.

▶ *Example 3*

Solving Equations Involving a Natural Logarithm

Solve for x.

    1. $\ln(x) = 6$ **2.** $5\ln(x + 1) = 22$ **3.** $\ln(2x) + 7 = 10.3$

Solution

1. We rewrite the expression $\ln(x) = 6$ in exponential form and compute using a calculator:

$$x = e^6 \approx 403.4288.$$

2.

$$5 \ln(x + 1) = 22 \qquad \text{Original equation.}$$
$$\ln(x + 1) = 4.4 \qquad \text{Divide both sides by 5.}$$
$$e^{4.4} = x + 1 \qquad \text{Write in exponential form.}$$
$$e^{4.4} - 1 = x \qquad \text{Subtract 1 from both sides.}$$
$$x \approx 80.4509 \qquad \text{Compute.}$$

We check that 80.4509 is an approximate solution to the original equation:

$$5 \ln(80.4509 + 1) = 5 \ln(81.4509) \approx 22.0000019 \approx 22$$

3.

$$\ln(2x) + 7 = 10.3 \qquad \text{Original equation.}$$
$$\ln(2x) = 3.3 \qquad \text{Subtract 7 from both sides.}$$
$$e^{3.3} = 2x \qquad \text{Write in exponential form.}$$
$$\frac{e^{3.3}}{2} = x \qquad \text{Divide both sides by 2.}$$
$$x \approx 13.5563 \qquad \text{Compute.}$$

We check that 13.5563 is an approximate solution to the original equation:

$$\ln(2(13.5563)) + 7 \approx 10.299999 \approx 10.3$$

To solve equations involving exponential terms with base e, we first simplify the equation (if necessary), then change to logarithmic form.

▶ Example 4

Solving an Exponential Equation with Base e

Solve for x.

1. $2e^x = 68$.   2. $1.5e^{3x} + 10 = 1393$

Solution

1.

$$2e^x = 68 \qquad \text{Original equation.}$$
$$e^x = 34 \qquad \text{Divide both sides by 2.}$$
$$\ln(34) = x \qquad \text{Write in logarithmic form.}$$
$$x \approx 3.5264 \qquad \text{Compute.}$$

We check that 3.5264 is an approximate solution to the original equation:

$$2e^{3.5264} \approx 68.0027 \approx 68$$

2.

$$1.5\,e^{3x} + 10 \;=\; 1393 \qquad\qquad \text{Original equation}$$

$$1.5\,e^{3x} \;=\; 1383 \qquad\qquad \text{Subtract 10 from both sides.}$$

$$e^{3x} \;=\; 922 \qquad\qquad \text{Divide both sides by 1.5.}$$

$$\ln(922) \;=\; 3x \qquad\qquad \text{Write in logarithmic form.}$$

$$\frac{\ln(922)}{3} \;=\; x \qquad\qquad \text{Divide both sides by 3.}$$

$$x \;\approx\; 2.2755 \qquad\qquad \text{Compute.}$$

We check that 2.2755 is an approximate solution to the original equation:

$$1.5\,e^{3(2.2755)} + 10 \;\approx\; 1392.9375 \;\approx\; 1393$$

Figure 3 shows the verification step on the calculator.

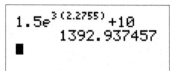

Figure 3. Verify your work.

Practice Set B

Solve for x. Round your answers to four decimal places. Turn the page to check your solutions.

7. $\ln(x) = 9.2$ **8.** $3\ln(x-2) = 18$ **9.** $e^x + 4 = 25$ **10.** $8\,e^{2x} = 436$

Practice Set A — Answers

1. $\ln(400) \approx 5.9915$
because $e^{5.9915} \approx 400$

2. $\ln(0.77) \approx -0.2615$
because $e^{-0.2615} \approx 0.77$

3. It is not possible to take the logarithm of a negative number in the set of real numbers.

4. $e^{\ln(38)} = 38$

5. $\ln(\sqrt[3]{e}) = \ln(e^{1/3}) = \frac{1}{3}$

6. $\ln(e^2) = 2$

C. EXPONENTIAL MODELS WITH BASE *e*

Exponential models with base *e* represent *continuous* exponential growth or decay. The constant *e* was named by the Swiss mathematician Leonhard Euler (1707–1783), who first investigated and discovered many of its properties. Scientists and mathematicians use the number *e* as a base for many real-world exponential models.

▶ *Example 5*

Using a Model with Base *e* to Make a Prediction

The population *P* (in thousands) of Bellevue, Washington, *t* years after 2000 is given by:

$$P = f(t) = 110\,e^{0.0198t}$$

   c.   Estimate the population of Bellevue in 2016.

   d.   When will the population of Bellevue reach 180 thousand?

Solution

a. For the year 2016, $t = 16$.

$$P = f(16) = 110\,e^{0.0198(16)} \approx 151$$

   We estimate the population of Bellevue in 2016 is 151 thousand.

b. Replace $f(t)$ with 180 (thousand) and solve for t.

$110\,e^{0.0198t} = 180$	$f(t) = 180$
$e^{0.0198t} = \dfrac{18}{11}$	Divide both sides by 110.
$\ln\left(\dfrac{18}{11}\right) = 0.0198t$	Write in logarithmic form.
$\dfrac{\ln\left(\frac{18}{11}\right)}{0.0198} = t$	Divide both sides by 0.0198.
$24.9 \approx t$	Compute.

The population of Bellevue, Washington, will reach 180 thousand in about 24.9 years, near the end of the year 2024.

▶ *Example 6*

Calculating Continuous Decay

Radon-222 decays at a *continuous* rate of 17.3% per day. This is modeled by the following equation where *A* represents the amount of Ra-222 remaining from a 100 mg sample after *t* days.

$$A = f(t) = 100\,e^{-0.0173t}$$

How long does it take for there to be 50 mg remaining in the sample?

Solution

Set the equation equal to 50 and solve for *t*.

$100\,e^{-0.0173t} = 50$	Substitute $f(t) = 50$.
$e^{-0.0173t} = 0.5$	Divide both sides by 100.
$\ln(0.5) = -0.0173t$	Rewrite in logarithmic form.
$\dfrac{\ln(0.5)}{-0.0173} = t$	Divide both sides by -0.0173
$t \approx 40.0663$	Compute.

In about 40 days, 50 mg of radon-222 will remain.

Because we began with 100 mg of Ra-222 and ended with 50 mg, this can be interpreted to mean that the *half-life* of radon-222 is about 40 days.

Exercises

Use a calculator to find the natural logarithm. Round your result to the fourth decimal place.

1. $\ln(10.83)$

2. $\ln(72,110)$

3. $\ln(4600)$

4. $\ln(0.25)$

Simplify mentally using the properties of logarithms. You can verify your result with a calculator.

5. $\ln(1)$

6. $\ln(e)$

7. $\ln(e^{12})$

8. $\ln(e^{-2})$

9. $\ln\left(\frac{1}{e^4}\right)$

10. $\ln\left(\frac{1}{e}\right)$

11. $\ln(\sqrt{e})$

12. $\ln(\sqrt[5]{e})$

13. $e^{\ln 9}$

14. $e^{\ln 3}$

Solve for *x*. Round your solutions to the fourth decimal place.

15. $\ln(x) = 4$

16. $\ln(x) = 7$

17. $9\ln(x-4) = 18$

18. $6\ln(x+3) = 24$

19. $\ln(6x) + 2 = 7$

20. $\ln(5x) - 8 = -2$

21. $4\,e^x = 530$

22. $8\,e^x = 936$

23. $2.75\,e^{x+3} = 17,050$

24. $8.4\,e^{x-2} = 2801.4$

25. $e^{2x-9} = 795.7$

26. $e^{4x+1} = 6213.5$

27. $9\,e^x - 240 = 6\,e^x + 270$

28. $8\,e^x - 733 = 6\,e^x + 422$

Practice Set B — Answers

7. $e^{9.2} \approx 9897.1291$

8. $e^6 + 2 \approx 405.4288$

9. $\ln(21) \approx 3.0445$

10. $\dfrac{\ln(54.5)}{2} \approx 1.9991$

29. A culture of bacteria grows according to the continuous growth model $B = f(t) = 500\,e^{0.062t}$ where B is the number of bacteria and t is in hours.

   a. Find and interpret $f(0)$.

   b. Find the number of bacteria after 6 hours.

   c. To the nearest tenth of an hour, determine how long it will take for the population to grow to 1200 bacteria.

30. A population of rabbits in a nature preserve grows according to the continuous growth model $R = f(t) = 8.8\,e^{0.052t}$ where R is the number of rabbits in thousands and t is in years.

   a. Find and interpret $f(0)$.

   b. Find the number of rabbits after 7 years.

   c. To the nearest tenth of a year, determine how long it will take for the population to grow to 50 thousand rabbits.

31. A person invests in account with interest compounded continuously according to the formula $A = f(t) = 15\,e^{0.044t}$ where A is in thousands of dollars and t is in years.

   a. Find and interpret $f(0)$.

   b. Find the value of the account after 10 years.

   c. To the nearest tenth of a year, determine how long it will take for the investment to grow to 40 thousand dollars.

32. A person invests in account with interest compounded continuously according to the formula $A = f(t) = 12\,e^{0.0375t}$ where A is in thousands of dollars and t is in years.

   a. Find and interpret $f(0)$.

   b. Find the value of the account after 8 years.

   c. To the nearest tenth of a year, how long it will take for the investment to grow to 25 thousand dollars?

33. A person buys a cup of coffee. The coffee's temperature, in $^\circ$F, is given by $f(t) = 132\,e^{-0.05t} + 68$ where t is in minutes. If the person waits until the coffee is 170° F before they begin to drink it, how long will they have to wait?

34. A person makes a cup of tea. The tea's temperature, in $^\circ$F, is given by $f(t) = 136\,e^{-0.05t} + 71$ where t is in minutes. If the person waits until the tea is 175°F before they begin to sip it, how long will they have to wait?

35. A medical lab has a 120 mg sample of Cobalt-60 which is a radioactive isotope of cobalt. The amount of Cobalt-60 (in milligrams) remaining in the sample after t years is given by $C = f(t) = 120\,e^{-0.1216t}$. How long will it take for the sample to have only 25 milligrams of Cobalt-60 remaining?

36. The fossil of the vertebrae of a whale was discovered by paleontologists and analyzed. They determined that the amount of Carbon-14 in the vertebrae at the time the whale was alive was approximately 16.5 milligrams. So the paleontologists knew that C the amount of C-14 (in mg) in the fossil would decay according to the equation $C = f(t) = 16.5\,e^{-0.00012t}$ where t is in years. How long ago did the whale live if there is only 0.04 mg of C-14 remaining?

37. The population P of a state can be modeled by the equation $P = f(t) = P_0\,e^{0.0145t}$ where P_0 is the population of the state now and t represents years from now. How long will it take the population to double?

38. The population P of a country can be modeled by the equation $P = f(t) = P_0\,e^{0.019t}$ where P_0 is the population of the state now and t represents years from now. How long will it take the population to triple?

CHAPTER 4

Quadratic Functions

A home run hit during a baseball game, a stone tossed off a cliff, a small rocket launched from the backyard — these events are all examples of projectile motion, which can be modeled with a quadratic function. In this chapter, we will study quadratic functions, their characteristics, and their applications to real world situations.

In the first section of this chapter, we will review some basic algebra skills, focusing on expanding and factoring polynomials. Then we'll define quadratic functions and represent these functions using tables, graphs, and equations. Finally, we will study how to calculate outputs, how to model data, and how to solve problems involving quadratic functions.

4.1 Expanding and Factoring Polynomials

OVERVIEW

This section examines algebraic expressions called polynomials. We have already used simple polynomials, for example $3x - 7$, when we worked with linear functions. We will practice multiplying and factoring polynomials as well as solving equations involving polynomials. It is likely you have studied the topics in this section in a previous algebra class and it is with that assumption we review them here. Refreshing your knowledge of some basic algebra skills will facilitate the study of quadratic functions, which is the focus of the rest of Chapter 4.

In this section, you will learn how to :

- ◆ Multiply polynomials.
- ◆ Use the FOIL method to multiply binomials.
- ◆ Identify and multiply special products.
- ◆ Factor polynomials with a greatest common factor.
- ◆ Factor trinomials.
- ◆ Apply the Zero Product Property.

A. MULTIPLICATION OF MONOMIALS AND POLYNOMIALS

We'll start with some review. The following are all examples of polynomials:

$$5x^4 - 6x^2 + 9x + 2$$
$$9x^2y - xy + y^2$$
$$4z^2 - 1$$

Recall that a **term** is a number, variable, or product of a number and one or more variables that are raised to powers. A **polynomial** or **polynomial expression** is the sum or difference of terms with the condition that all the exponents on the variables must be nonnegative integers. The first polynomial above contains four terms.

The term in a polynomial that has the largest combined powers on its exponents is called the **leading term**. In all three examples above, the leading term has been written first.

The numerical factor of a term — for example, the 5 in $5x^4$ — is called a **coefficient**. The coefficient on the term y^2 is 1 because y^2 can be written as $1 \cdot y^2$. Coefficients can be positive, negative, or even zero, and they can be whole numbers, decimals, or fractions. The coefficient of the leading term is called the **leading coefficient**.

If a term in a polynomial does not contain a variable, it's called a **constant**. A polynomial containing only one term, such as $5x^4$, is called a **monomial**. A polynomial containing two terms, such as $2x - 9$, is called a **binomial**. A polynomial containing three terms, such as $-3x^2 + 8x + 7$, is called a **trinomial**.

When we studied the product property for exponents in Section 2.1, we looked at the product of two monomials. Recall that the product property for exponents states that $x^m x^n = x^{m+n}$.

▶ Example 1

Multiplying Monomials

Simplify.

1. $(9x^4y^2)(-4x^5y^3)$ **2.** $(-2a^3bc^6)(-8ab^7c^2)$

Solution

1. $(9x^4y^2)(-4x^5y^3)$

$$
\begin{aligned}
(9x^4y^2)(-4x^5y^3) &= (9)(-4)(x^4x^5)(y^2y^3) && \text{Regroup and reorder factors.} \\
&= -36x^{4+5}y^{2+3} && \text{Apply the product property for exponents.} \\
&= -36x^9y^5 && \text{Simplify.}
\end{aligned}
$$

2. $(-2a^3bc^6)(-8ab^7c^2)$

$$
\begin{aligned}
(-2a^3bc^6)(-8ab^7c^2) &= (-2)(-8)(a^3a)(bb^7)(c^6c^2) && \text{Regroup and reorder factors.} \\
&= 16a^{3+1}b^{1+7}c^{6+2} && \text{Apply the product property for exponents.} \\
&= 16a^4b^8c^8 && \text{Simplify.}
\end{aligned}
$$

Multiplying a monomial by a polynomial requires us to use the distributive property of multiplication over addition and subtraction. In this case, we multiply each term in the polynomial by the monomial and simplify.

▶ Example 2

Multiplying a Monomial by a Polynomial

Multiply.

1. $3x^2(7x^5 - 4x^3 + 6)$ **2.** $(2x^2 + 11xy - 3y^2)(5xy^2)$

Solution

1. $3x^2(7x^5 - 4x^3 + 6)$

$$
\begin{aligned}
3x^2(7x^5 - 4x^3 + 6) &= 3x^2(7x^5) - 3x^2(4x^3) + 3x^2(6) && \text{Apply the distributive property.} \\
&= 21(x^2x^5) - 12(x^2x^3) + 18x^2 && \text{Regroup and reorder factors. Multiply} \\
& && \text{coefficients.} \\
&= 21x^7 - 12x^5 + 18x^2 && \text{Simplify each term, using the product} \\
& && \text{property for exponents.}
\end{aligned}
$$

2. $2x^2 + 11xy - 3y^2)(5xy^2)$

$(2x^2 + 11xy - 3y^2)(5xy^2)$ = $2x^2(5xy^2) + 11xy(5xy^2) - 3y^2(5xy^2)$ Apply the distributive property.

= $10(x^2x)(y^2) + 55(xx)(yy^2) - 15x(y^2y^2)$ Regroup and reorder factors. Multiply coefficients.

= $10x^3y^2 + 55x^2y^3 - 15xy^4$ Simplify each term, using the product property for exponents.

Notice in Example 2 that it doesn't matter if the monomial term is to the left or the right of the other polynomial. The distributive property works the same way in either case.

Practice Set A

Now it's your turn to multiply with monomial factors. When you're finished, turn the page to check your work.

1. $(-7x^5y^3)(6x^2y)$
2. $(11a^9b^2c^3)(2ab^4c)$
3. $(-3x^2)(3x^2 + 8x - 4)$
4. $(4x^3y - 9x^2y^2 + y)(5y)$
5. $(8xy^2)(2x^3y^3 + 3xy - 1)$

B. MULTIPLYING BINOMIALS

So far we have been working with polynomials in more than one variable. For the remainder of this section we will primarily focus on polynomials in a single variable as we build skills useful in working with quadratic functions.

When multiplying two binomials, we sometimes use a process called the FOIL method to find the product. It's called "FOIL" because we multiply the **First** terms, the **Outer** terms, the **Inner terms**, and then the **Last terms** of each binomial.

The FOIL method arises out of the distributive property. We simply multiply each term of the first binomial by each term of the second binomial and then combine like terms. Recall like terms have identical variable factors, including exponents.

first terms
last terms

$$(ax + b)(cx + d) = acx^2 + adx + bcx + bd$$

inner terms

outer terms

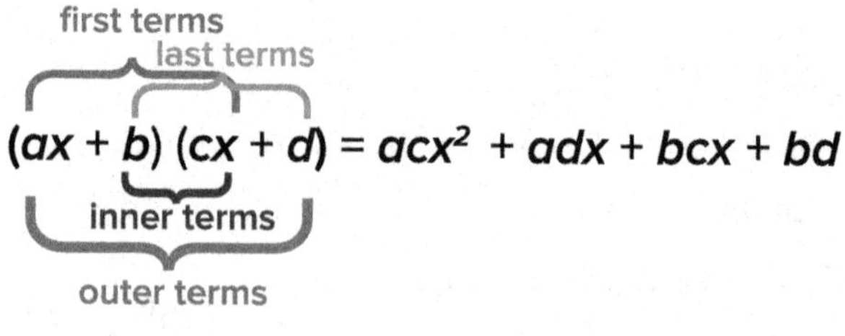

Figure 1. The FOIL method.

Given two binomials, then, we can use FOIL to simplify the expression by following these steps:

1. Multiply the first terms of each binomial.

2. Multiply the outer terms of the binomials.

3. Multiply the inner terms of the binomials.

4. Multiply the last terms of each binomial.

5. Add the products obtqined in steps 1–4.

6. Combine like terms and simplify.

▶ Example 3

Using FOIL to Multiply Binomials

Use FOIL to find the product: $(2x - 10)(3x + 3)$

Solution

Find the product of the first terms (Figure 2).
Find the product of the outer terms (Figure 3).
Find the product of the inner terms (Figure 4).
Find the product of the last terms (Figure 5).

$6x^2 + 6x - 54x - 54$ Add the products.

$6x^2 + (6x - 54x) - 54$ Group like terms.

$6x^2 - 48x - 54$ Combine like terms.

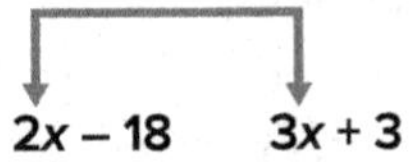

2x – 18   3x + 3   **2x · 3x = 6x²**

Figure 2. First terms.

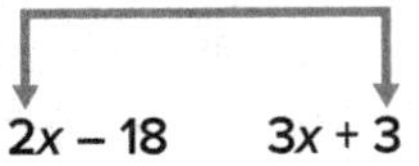

2x – 18   3x + 3   **2x · 3 = 6x**

Figure 3. Outer terms.

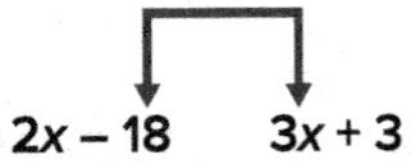

2x – 18   3x + 3   **–18 · 3x = –54x**

Figure 4. Inner terms.

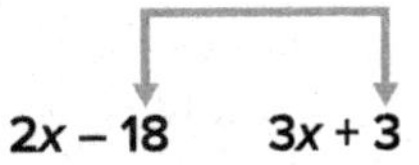

2x – 18   3x + 3   **–18 · 3 = –54**

Figure 5. Last terms.

Let's look at a few more examples of multiplying binomials pairs. We'll use binomials of form, $x \pm a$, and give attention to the pattern in the products. Specifically, watch how we find the middle term and the last term of the resulting trinomial.

▶ Example 4

Multiplying Binomials

Use the FOIL method to multiply each pair of binomials.

  1. $(x + 9)(x + 2)$ **2.** $(x + 7)(x - 3)$ **3.** $(x - 6)(x + 1)$

Solution

1. $(x + 9)(x + 2)$

$$\begin{aligned} (x + 9)(x + 2) &= x^2 + 2x + 9x + 18 \qquad \text{Use the FOIL method.} \\ &= x^2 + 11x + 18 \qquad \text{Combine like terms.} \end{aligned}$$

2. $(x + 7)(x - 3)$

$$\begin{aligned} (x + 7)(x - 3) &= x^2 - 3x + 7x - 21 \qquad \text{Use the FOIL method.} \\ &= x^2 + 4x - 21 \qquad \text{Combine like terms.} \end{aligned}$$

3. $(x - 6)(x + 1)$

$$(x - 6)(x + 1) = x^2 + x - 6x - 6 \quad \text{Use the FOIL method.}$$
$$= x^2 - 5x - 6 \quad \text{Combine like terms.}$$

We can see from Example 4 that the coefficient of the middle term of the trinomial is the *sum* of the original constants. The last term of the trinomial is the *product* of the original constants. We can use this observation as a two-part shortcut to finding products of binomials. Perform the step of combining the outer and inner products mentally:

$$(x - 5)(x + 8) = x^2 + 3x - 40 \quad \text{Expand and mentally combine the outer and inner products:}$$
$$-5x + 8x = 3x.$$

Be careful, though! If the binomials are not of form $x \pm a$, the mental math used to find the middle term is more involved:

$$(2x + 5)(3x - 4) = 6x^2 + 7x - 20 \quad \text{Expand and mentally combine the outer and inner products:}$$
$$-8x + 15x = 7x.$$

If a product of two binomials looks challenging, you can always write down all four products using the FOIL method before combining like terms. The FOIL method is a dependable way to keep track of the process.

Practice Set B

Now it's your turn to multiply binomials. When you're finished, turn the page to check your work.

6. $(x + 2)(x + 7)$ **8.** $(x - 8)(x + 1)$ **10.** $(5x + 4)(2x + 1)$

7. $(x + 3)(x - 9)$ **9.** $(a - 10b)(a - 6b)$ **11.** $(3x^2 - x)(4x + 7)$

Practice Set A — Answers

1. $-42x^7 y^4$

2. $22 a^{10} b^6 c^4$

3. $-9x^4 - 24x^3 - 12x^2$

4. $20x^3 y^2 - 45x^2 y^3 + 5y^2$

5. $16x^4 y^5 + 24x^2 y^3 - 8xy^2$

C. SPECIAL BINOMIAL PRODUCTS

Perfect Square Trinomials

Certain binomial products have special forms. When a binomial is squared, the result is called a **perfect square trinomial**. We can find the square by multiplying the binomial by itself. However, there's a special form that each of these perfect square trinomials takes, and learning the form makes squaring binomials easier and faster.

> **Perfect Square Trinomials**
>
> $$(x + a)^2 = (x + a)(x + a)$$
> $$= x^2 + 2ax + a^2$$

Let's look at a few perfect square trinomials to familiarize ourselves with the form.

$$(x + 5)^2 = (x + 5)(x + 5) = x^2 + 10x + 25$$
$$(x - 3)^2 = (x - 3)(x - 3) = x^2 - 6x + 9$$
$$(4x - 1)^2 = (4x - 1)(4x - 1) = 16x^2 - 8x + 1$$

Notice that the first term of each trinomial is the square of the first term of the binomial. Similarly, the last term of each trinomial is the square of the last term of the binomial. The middle term is double the product of the two binomial terms. Last, we see that the sign of the middle term of the trinomial is the same as the sign between the terms of the binomial.

The square of a binomial is the first term squared, plus double the product of the two terms, plus the second term squared.

Given a binomial, we square it using the formula for perfect square trinomials by following these steps.

1. Square the first term of the binomial.

2. Square the last term of the binomial.

3. For the middle term of the trinomial, double the product of the two binomial terms.

4. Add and simplify.

▶ *Example 5*

Squaring a Binomial

Expand $(3x - 8)^2$.

Solution

$$(3x)^2 - 2(3x)(8) + (-8)^2$$ Begin by squaring the first term and the last term. For the middle term of the trinomial, double the product of the two terms.

$$9x^2 - 48x + 64$$ Simplify.

Difference of Squares

Another special product occurs when we multiply a binomial by another binomial with the same terms but the opposite sign.

Let's see what happens when we expand $(x + 4)(x - 4)$.

$$(x + 4)(x - 4) = x^2 - 4x + 4x - 16$$
$$= x^2 - 16$$

> ### Difference of Squares
>
> When a binomial is multiplied by a binomial with the same terms separated by the opposite sign, the result is the square of the first term minus the square of the last term.
>
> $$(a + b)(a - b) = a^2 - b^2$$

The middle term is the sum of opposites, in this case $-4x$ and $4x$. The sum is $0x = 0$. So in effect, the middle term drops out, resulting in a **difference of squares**. We can better understand this description by writing the solution as

$$x^2 - 16 = (x)^2 - (4)^2$$

The second form emphasizes we are subtracting (finding the difference of) two expressions that have been squared.

Just as we did with the perfect square trinomials, let's look at a few examples.

$$(x + 5)(x - 5) = x^2 - 25$$
$$(x + 11)(x - 11) = x^2 - 121$$
$$(2x + 3)(2x - 3) = 4x^2 - 9$$

Because the sign is opposite in the second binomial, the outer and inner terms add to zero, and we are left only with the square of the first term minus the square of the last term, a difference of squares.

Given a binomial multiplied by a binomial with the same terms but the opposite sign, we can find the difference of squares by following these steps.

1. Square the first term of the binomials.

2. Square the last term of the binomials.

3. Subtract the square of the last term from the square of the first term.

Practice Set B — Answers

6. $x^2 + 9x + 14$ **8.** $x^2 - 7x - 8$ **10.** $10x^2 + 13x + 4$

7. $x^2 - 6x - 27$ **9.** $a^2 - 16ab + 60b^2$ **11.** $12x^3 + 17x^2 - 7x$

▶ *Example 6*

Multiplying Binomials Resulting in a Difference of Squares

Multiply.

   1. $(x + 7)(x - 7)$ **2.** $(9x + 4)(9x - 4)$

Solution

1. Square the first term: $(x)^2 = x^2$. Square the last term: $7^2 = 49$. Write the difference: $x^2 - 49$.

2. Square the first term: $(9x)^2 = 81\,x^2$. Square the last term: $4^2 = 16$. Write the difference: $81\,x^2 - 16$.

Practice Set C

Now it's your turn. When you are finished with these problems, turn the page to check your work. Expand the following.

  12. $(x + 3)^2$ **13.** $(x - 8)^2$ **14.** $(2x + 11)^2$

Multiply the following.

  15. $(x + 1)(x - 1)$ **16.** $(x + 12)(x - 12)$ **17.** $(2x + 7)(2x - 7)$

D. FACTORING POLYNOMIALS AND THE GREATEST COMMON FACTOR

Many polynomial expressions can be written in more useful forms by factoring. Factoring involves finding polynomial expressions that can be multiplied together to give the original polynomial. We have a variety of methods for factoring polynomial expressions. One way to think of factoring is to see it as "undoing" the multiplying and expanding that we just practiced.

When we study fractions, we learn that the **greatest common factor** (GCF) of two numbers is the largest number that divides both numbers evenly. For example, 4 is the GCF of 16 and 20 because it's the largest number that divides both 16 and 20 evenly. The GCF of polynomials works the same way. $4x$ is the GCF of $16x$ and $20x^2$ because

> ### Greatest Common Factor
>
> The **greatest common factor** (GCF) of polynomials is the largest polynomial that divides evenly into the polynomials.

it's the largest polynomial that divides both $16x$ and $20\,x^2$. When factoring a polynomial expression, our first step is to check for a GCF. Look for the GCF of the coefficients, and then look for the GCF of the variables.

Given a polynomial expression, we factor out the greatest common factor by following these steps.

  1. Identify the GCF of the coefficients.

  2. Identify the GCF of the variables.

  3. Combine to find the GCF of the polynomial.

4. Use the distributive property to factor the GCF out of the polynomial. Determine what the GCF needs to be multiplied by to obtain each term in the original polynomial expression.

5. Verify your work by multiplying out your factored result.

▶ Example 7

Factoring the Greatest Common Factor

Factor the polynomials.

    1. $14x^2 + 21x$ **2.** $6x^4 - 9x^3 + 12x^2$

Solution

1. The GCF of the coefficients is 7 because 7 is the largest integer that divides 14 and 21 evenly. The GCF of the variables is x, because x is the largest power of the variable that divides x^2 and x. Therefore, the GFC of $14x^2 + 21x$ is $7x$.

    We ask, "What is $7x$ multiplied by to obtain the terms of the original expression?" Then we use the distributive property: $7x(?? + ?) = 14x^2 + 21x$ The answer is $7x(2x + 3)$ By mentally multiplying the factors, we see that we do get back the original polynomial, so this solution checks out.

2. The GCF of the coefficients 6, 9, and 12 is 3. The GCF of the variable expressions x^4, x^3, and x^2 is x^2. So the GCF of the expression is $3x^2$. Next, we use the distributive property to find the answer: $6x^4 - 9x^3 + 12x^2 = 3x^2(2x^2 - 3x + 4)$

▶ Example 8

Factoring the Greatest Common Factor with Two Variables

Factor $6x^3y^3 + 45x^2y^2 + 21xy$.

Solution

The GCF of the coefficients 6, 45, and 21 is 3. The GCF of x^3, x^2, and x is x. The GCF of y^3, y^2, and y is y. Combine these to find the GCF of the polynomial, $3xy$. Next, determine what the GCF needs to be multiplied by to obtain each term of the polynomial.

    We find that $3xy(2x^2y^2) = 6x^3y^3$, $3xy(15xy) = 45x^2y^2$, and $3xy(7) = 21xy$. Finally, write the factored expression as the product of the GCF and the sum of the terms we needed to multiply by:

$$3xy(2x^2y^2 + 15xy + 7)$$

    After factoring, we should always at least mentally check our work by multiplying. Use the distributive property to confirm that $(3xy)(2x^2y^2 + 15xy + 7) = 6x^3y^3 + 45x^2y^2 + 21xy$. The GCF of two or more expressions each in the form x^n will always be the exponent of lowest degree.

Practice Set C — Answers

12. $x^2 + 6x + 9$	**14.** $4x^2 + 44x + 121$	**16.** $x^2 - 144$
13. $x^2 - 16x - 64$	**15.** $x^2 - 1$	**17.** $4x^2 - 49$

Practice Set D

Factor each polynomial using the greatest common factor of the terms. When you're finished, turn the page to check your work.

18. $24x^2 + 80x$

19. $5x^3 - 35x^2 + 10x$

20. $8a^6 - 10a^4 + 6a^3$

21. $12x^3y^2 + 3x^2y^3 + 18xy^4$

E. FACTORING TRINOMIALS

Factoring a Trinomial with Leading Coefficient 1

Although we should begin by looking for a GCF, pulling out the GCF is not the only way that many polynomial expressions can be factored. The polynomial $x^2 + 5x + 6$ has a GCF of 1, but it can be written as the product of the factors $(x + 2)$ and $(x + 3)$. We can check this using the FOIL method to multiply:

$$(x + 2)(x + 3) = x^2 + 5x + 6$$

> ### Factoring a Trinomial with Leading Coefficient 1
>
> A non-prime trinomial of the form $x^2 + bx + c$ can be written in factored form as $(x + p)(x + q)$ where $pq = c$ and $p + q = b$.

As we see above, trinomials of the form $x^2 + bx + c$ can sometimes be rewritten as the product of two binomials. Finding the constants for the binomial factors involves finding two numbers having a product of c and a sum of b. Considering the trinomial $x^2 + 10x + 16$, for example, 2 and 8 are two numbers that have a product of 16 and a sum of 10. We then rewrite the trinomial as the product of $(x + 2)$ and $(x + 8)$:

$$x^2 + 10x + 16 = (x + 2)(x + 8)$$

Not every trinomial can be factored as a product of binomials. Some polynomials cannot be factored. These polynomials are said to be prime.

Given a trinomial in the form $x^2 + bx + c$, we factor it by following these steps:

1. List factors of c.

2. Find p and q, a pair of factors of c with a sum of b.

3. Write the factored expression $(x + p)(x + q)$.

Factors of −15	Sum of Factors
1, −15	−14
−1, 15	14
3, −5	−2
−3, 5	2

Figure 6.

▶ *Example 9*

Factoring a Trinomial with Leading Coefficient 1

Factor $x^2 + 2x - 15$.

Solution

We have a trinomial with leading coefficient 1, $b = 2$, and $c = -15$. We need to find two numbers with a product of -15 and a sum of 2. In Figure 6, we list factors until we find a pair with the desired sum. Once we have identified p and q as -3 and 5, we write the factored form as $(x - 3)(x + 5)$. We can check our work by multiplying. Expand to confirm that $(x - 3)(x + 5) = x^2 + 2x - 15$.

The order in which we write the binomial factors doesn't matter, by the way, because multiplication is commutative, meaning that $a \cdot b = b \cdot a$.

▶ *Example 10*

Factoring a Trinomial with Leading Coefficient 1

Write each polynomi in factored form.

1. $x^2 - 4x - 21$ 2. $x^2 + 9x + 8$ 3. $x^2 - 13x + 30$ 4. $x^2 + 5x - 2$

Solution

1. We look for a pair of numbers whose product is –21 and whose sum is –4. We find $(-7)(3) = -21$ and $-7 + 3 = -4$. So the factors of $x^2 - 4x - 21$ are $(x - 7)$ and $(x + 3)$.
   Now we check our work: $(x - 7)(x + 3) = x^2 - 4x - 21$

2. We look for a pair of numbers whose product is 8 and whose sum is 9. We find $(8)(1) = 8$ and $8 + 1 = 9$. So the factors of $x^2 + 9x + 8$ are $(x + 8)$ and $(x + 1)$.
   We check our work: $(x + 8)(x + 1) = x^2 + 9x + 8$

3. We look for a pair of numbers whose product is 30 and whose sum is –13. We find $(-10)(-3) = 30$ and $-10 + -3 = -13$. So the factors of $x^2 - 13x + 30$ are $(x - 10)$ and $(x - 3)$.
   Check: $(x - 10)(x - 3) = x^2 - 13x + 30$

4. We look for a pair of numbers whose product is –2 and whose sum is 5. We find there are no integers that satisfy these conditions. So $x^2 + 5x - 2$ is a prime polynomial and cannot be factored.

Some trinomials have a greatest common factor that can be pulled out before we try to find two binomial factors. And some trinomials have more than one variable. We apply what we know about factoring to these situations.

▶ *Example 11*

Factoring Trinomials

1.   Factor: $4x^3 + 12x^2 - 160x$ 2.   Factor: $x^2 + 10xy + 24y^2$

Solution

1. The GCF of the coefficients is 4, and the GCF of the variables is x. That means that $4x^3 + 12x^2 - 160x = 4x(x^2 + 3x - 40)$. Next we look for a pair of numbers whose product is –40 and whose sum is 3. We find that 8 and –5 will satisfy this condition. So the complete factorization of the polynomial is $4x^3 + 12x^2 - 160x = 4x(x + 8)(x - 5)$.

Practice Set D — Answers

18. $8x(3x + 10)$	**20.** $2a^3(4a^3 - 5a + 3)$
19. $5x(x^2 - 7x + 2)$	**21.** $3xy^2(4x^2 + xy + 6y^2)$

2. There is no GCF of the terms other than 1, so we look for a pair of numbers whose product is 24 and whose sum is 10. We find that 6 and 4 satisfy this condition. Now, taking into account both variables in the polynomial, we find that $x^2 + 10xy + 24y^2 = (x + 6y)(x + 4y)$.

Factoring a Trinomial when the Leading Coefficient is not 1

We turn our attention to trinomials with leading coefficients other than 1 and with no greatest common factor other than 1. These are slightly more complicated to factor.

Suppose we want to factor $2x^2 - 7x - 15$. If it is factorable into two binomials, the product of the first terms will be $2x^2$, and the product of the last terms will be -15. We use trial and error and list some of the possible factors along with their products in Figure 7.

We have not listed all of the possible factors, but we find that the fourth trial gives the correct middle term. The factors of $2x^2 - 7x - 15$ are $(2x + 3)$ and $(x - 5)$. That means that $2x^2 - 7x - 15 = (2x + 3)(x - 5)$.

Possible Factors	First Term	Middle Term	Last Term
$(2x - 1)(x + 15)$	$2x^2$	$30x - x = 29x$	-15
$(2x - 15)(x + 1)$	$2x^2$	$2x - 15x = -13x$	-15
$(2x - 3)(x + 5)$	$2x^2$	$10x - 3x = 7x$	-15
$(2x + 3)(x - 5)$	$2x^2$	$-10x + 3x = -7x$	-15

Figure 7.

When using trial and error to solve a problem, it helps to pay attention to the *incorrect* results during the process. Observing that the middle term is the wrong sign — positive vs. negative— or too large or too small can help guide the next trial. In this way, we can shorten the process of finding the correct factorization.

▶ *Example 12*

Factoring a Trinomial by Trial and Error

Factor $5x^2 + 13x - 6$.

Solution

If this trinomial factors into two binomials, the product of the first terms will be $5x^2$ and the product of the last terms will be -6. In Figure 8, we list possible factors along with their products until we find the correct pair.

We can check our work by multiplying.

Possible Factors	First Term	Middle Term	Last Term	Observation
$(5x + 1)(x - 6)$	$5x^2$	$-30x + x = -29x$	-6	The middle term should be positive.
$(5x + 6)(x - 1)$	$5x^2$	$-5x + 6x = x$	-6	The middle term is positive but too small.
$(5x + 3)(x - 2)$	$5x^2$	$-10x + 3x = -7x$	-6	The middle term should be positive.
$(5x - 2)(x + 3)$	$5x^2$	$15x - 2x = 13x$	-6	This gives the correct middle term.

Figure 8.

Expand to confirm that $(5x - 2)(x + 3) = 5x^2 + 13x - 6$.

Factoring a Perfect Square Trinomial

Recall that a perfect square trinomial is a trinomial that can be written as the square of a binomial. The square of a binomial is the first term squared, plus double the product of the two terms, plus the second term squared:

$$a^2 + 2ab + b^2 = (a + b)^2 \text{ and}$$

$$a^2 - 2ab + b^2 = (a - b)^2$$

We can use the preceding equations to factor any perfect square trinomial. Given a perfect square trinomial, then, we factor it into the square of a binomial by following these steps.

1. Confirm that the first and last term are perfect squares.

2. Confirm that the middle term is twice the product of a and b.

3. Write the factored form as $(a + b)^2$.

▶ *Example 13*

Factoring a Perfect Square Trinomial

Factor $25x^2 + 20x + 4$.

Solution

Notice that $25x^2$ and 4 are perfect squares because $25x^2 = (5x)^2$ and $4 = 2^2$. Now check to see if the middle term is twice the product of $5x$ and 2. The middle term is, indeed, twice the product: $2(5x)(2) = 20x$. Therefore, the trinomial is a perfect square trinomial and can be written as $(5x + 2)^2$.
Check by expanding the solution: $(5x + 2)^2 = (5x + 2)(5x + 2) = 25x^2 + 20x + 4$

Factoring a Difference of Squares

A difference of squares is a perfect square subtracted from a perfect square. Recall that a difference of squares can be rewritten as factors containing the same terms but opposite signs because the middle terms add to zero when the two factors are multiplied:

$$a^2 - b^2 = (a + b)(a - b)$$

We can use this equation to factor any differences of squares. Given a difference of squares, we factor it into binomials by following these steps.

1. Confirm that the first and last term are perfect squares.

2. Write the factored form as $(a + b)(a - b)$.

▶ *Example 14*

Factoring a Difference of Squares

    1.   Factor $x^2 - 36$.   **2.**   Factor $9x^2 - 25$

Solution

1. Notice that x^2 and 36 are perfect squares because $x^2 = (x)^2$ and $36 = 6^2$. So the factored form of the difference of squares $x^2 - 36$ is $(x + 6)(x - 6)$.

2. Notice that $9x^2$ and 25 are perfect squares because $9x^2 = (3x)^2$ and $25 = 5^2$. So the factored form of the difference of squares $9x^2 - 25$ is $(3x + 5)(3x - 5)$.

Practice Set E

Now it's time for you to do some factoring. When you're finished, turn the page to check your work.

22. $x^2 + 7x + 12$	**25.** $6x^2 + x - 1$	**28.** $81x^2 - 100$
23. $x^2 - 6x - 16$	**26.** $49x^2 - 14x + 1$.	
24. $2x^2 + 9x + 9$	**27.** $x^2 - 4$	

F. ZERO PRODUCT PROPERTY

A **quadratic equation** is an equation in one variable that contains a polynomial having 2 as the highest power on the variable. For example, equations such as $2x^2 + 3x + 1 = 0$ and $x^2 - 4 = 0$ are quadratic equations. They are used in many ways in the fields of engineering, architecture, finance, biological science, and, of course, mathematics.

    Sometimes the easiest method of solving a quadratic equation involves factoring. Factoring involves finding polynomial expressions that can be multiplied together to give the original polynomial. When we factor, we change a single polynomial that is the sum of terms into an expression that is the product of polynomials. If a quadratic expression can be factored, we write it as a product of linear expressions. For example, the quadratic equation $2x^2 + 3x + 1 = 0$ can be written in factored form as $(2x + 1)(x + 1) = 0$.

    Solving by factoring depends on the **zero-product property**, which states that if $a \bullet b = 0$, then $a = 0$ or $b = 0$, where a and b are real numbers or algebraic expressions. In other words, if the product of two numbers or two expressions equals zero, then at least one of the numbers or one of the expressions *must* equal zero.

    Given a quadratic equation that has a factorable expression when set equal to 0, we solve it by following these steps:

    1.   Set the equation equal to zero (if it's not already).

    2.   Factor the polynomial.

    3.   Solve using the zero-product property by setting each factor equal to zero and solving for the variable.

> **The Zero-Product Property and Quadratic Equations**
>
> The **zero-product property** states that if $a \bullet b = 0$, then $a = 0$ or $b = 0$, where a and b are real numbers or algebraic expressions.

▶ *Example 15*

Solving a Quadratic Equation by Factoring

Solve the equation by factoring $x^2 + x - 6 = 0$.

Solution

To factor the left-hand side, we look for two numbers whose product equals −6 and whose sum equals 1. Note that only one pair of numbers will work: 3 and −2. Next, we rewrite the equation:

$$(x - 2)(x + 3) = 0$$

By the zero-product property one or both of these factors must be 0. We set each factor equal to zero and solve.

$$x - 2 = 0 \quad \text{or} \quad x + 3 = 0$$
$$x = 2 \quad \text{or} \quad x = -3$$

We can check our solutions in the original equation as follows:

Check for $x = 2$: $(2)^2 + 2 - 6 \ = \ 4 + 2 - 6 \ = \ 0$

Check for $x = -3$: $(-3)^2 + -3 - 6 \ = \ 9 - 3 - 6 \ = \ 0$

In both cases we get a true statement, which means that both 2 and −3 are solutions to the original equation.

▶ *Example 16*

Solving a Quadratic Equation by Factoring

Solve the quadratic equations by factoring. Check your solution(s).

   1. $2x^2 + 11x - 6 \ = \ 0$ **2.** $x^2 - 8x + 16 \ = \ 0$ **3.** $x^2 - 144 \ = \ 0$

Solution

1. We can factor the polynomial by trial and error. We find that $2x^2 + 11x - 6 \ = \ 0$ becomes $(2x - 1)(x + 6) \ = \ 0$. By the zero-product property one or both of these factors must be 0. We set each factor equal to zero and solve.

$$2x - 1 = 0 \quad \text{or} \quad x + 6 = 0$$
$$2x = 1 \quad \text{or} \quad x = -6$$
$$x = \frac{1}{2}$$

Practice Set E — Answers

22. $(x + 3)(x + 4)$ **24.** $(2x + 3)(x + 3)$ **26.** $(7x - 1)^2$ **28.** $(9x + 10)(9x - 10)$

23. $(x - 8)(x + 2)$ **25.** $(3x - 1)(2x + 1)$ **27.** $(x + 4)(x - 4)$

We can use a calculator to check our solutions in the original equation:

Check for $x = \frac{1}{2}$: $2\left(\frac{1}{2}\right)^2 + 11\left(\frac{1}{2}\right) - 6 = 0$

Check for $x = -6$: $2(-6)^2 + 11(-6) - 6 = 0$

So $\frac{1}{2}$ and -6 are solutions to the original equation.

2. We can factor the polynomial $x^2 - 8x + 16 = 0$ and use the zero product rule to solve.

$(x - 4)(x - 4) = 0$	This is a perfect square trinomial.
$(x - 4) = 0$	Each factor gives the same equation.
$x = 4$	There is only one solution to the equation.

The solution is $x = 4$. Remember to check the solution on your calculator.

3. The polynomial in the equation $x^2 - 144 = 0$ has the form of the difference of squares. It factors to give $(x + 12)(x - 12) = 0$. By the zero product property, then:

$x + 12 = 0$ or $x - 12 = 0$

$x = 12$ or $x = -12$

Check both solutions on your calculator.

Some single variable polynomial equations have a greatest common factor (GCF) and we can use the distributive property to rewrite the polynomial as a product having the GCF as a factor. If the GCF is a monomial containing the variable, then one of the solutions is always 0.

▶ *Example 15*

Using the Zero-Product Property to Solve a Polynomial Equation with a GCF.

Solve for x.

1. $4x^2 - 28x = 0$

2. $x^3 + 14x^2 + 45x = 0$

Solution

1. The GCF of the polynomial is $4x$, so the equation can be written in factor form:

$4x(x - 7) = 0$

Using the zero product property, we know:

$4x = 0$ or $x - 7 = 0$

$x = 0$ or $x = 7$

2. The GCF of the polynomial is x so the equation can be written in factor form:

$$x(x^2 + 14x + 45) = 0$$

We factor the trinomial further into two binomials: $x(x + 5)(x + 9) = 0$.
Expanding the zero product property to a product of three factors we know:

$$x = 0 \quad \text{or} \quad x + 5 = 0 \quad \text{or} \quad x + 9 = 0$$

$$x = 0 \quad \text{or} \quad x = -5 \quad \text{or} \quad x = -9$$

You can check all three solutions in the original equation.

Practice Set F

Now it's your turn. Solve the following equations by factoring. When finished, check your work on the next page.

29. $x^2 - 5x - 6 = 0$

30. $x^2 - 4x - 21 = 0$

31. $x^2 - 25 = 0$

32. $x^3 + 11x^2 + 10x = 0$

Exercises

For the following exercises, find the product.

1. $(6x^2 y^7)(8x^5 y)$

2. $(-3x^8 y^2)(9xy^3)$

3. $(-7a^4 b)(2ab^6)$

4. $(-2ab^3)(-a^2 b)$

5. $(4x)(x^2 + 9x - 5)$

6. $(8x)(2x^2 - 3x + 1)$

7. $(3x^2)(-7x^2 + 11x - 2)$

8. $(10x^2)(-6x^2 + 3x - 8)$

9. $(6a^2 b - 5ab + 4b)(-3ab)$

10. $(-5ab)(3a^2 b - 8ab + 2b)$

11. $(2x^2 y)(x^2 + 9xy - y^2)$

12. $(x^2 + 2xy + y^2)(7xy^2)$

Multiply the binomials.

13. $(x + 2)(x - 4)$

14. $(x + 4)(x - 3)$

15. $(x + 11)(x + 8)$

16. $(x + 12)(x + 2)$

17. $(x - 9)(x - 6)$

18. $(x - 7)(x - 5)$

19. $(x^2 - 4)(x + 8)$

20. $(x^2 - 6)(x + 7)$

21. $(8a - 4)(a^3 + 3a)$

22. $(3d^2 - 5d)(2d^2 + 9)$

23. $(5x^3 + x^2)(4x^2 - 3)$

24. $(6x^3 + 1)(2x^2 - 7x)$

25. $(4a + b)(3a - 5b)$

26. $(2a - 11b)(5a - 4b)$

For the following exercises, expand the squared binomial.

27. $(x + 9)^2$

28. $(y - 10)^2$

29. $(a - 7)^2$

30. $(x + 11)^2$

31. $(2m - 3)^2$

32. $(3a + 5)^2$

For the following exercises, multiply the binomials.

33. $(x + 2)(x - 2)$

34. $(a - 8)(a + 8)$

35. $(b - 6)(b + 6)$

36. $(x + 15)(x - 15)$

37. $(4x + 1)(4x - 1)$

38. $(11x - 3)(11x + 3)$

39. $(2a + 7b)(2a - 7b)$

40. $(3a + 4b)(3a - 4b)$

For the following exercises, find the greatest common factor.

41. $14x + 4xy - 18xy^2$

42. $49mb^2 - 35m^2b + 77m$

43. $30x^3y - 45x^2y^2 + 135xy^3$

44. $200p^3m^3 - 30p^2m^3 + 40m^3$

45. $36j^4k^2 - 18j^3k^3 + 54j^2k^4$

46. $6y^4 - 2y^3 + 3y^2 - y$

47. $80r^7 - 16r^3t$

48. $44x^6y - 55x^5y^2$

For the following exercises, factor the trinomial.

49. $a^2 + 9a - 22$

50. $a^2 + 5a - 24$

51. $x^2 + 12x + 35$

52. $x^2 + 10x + 24$

53. $x^2 - 9x + 8$

54. $x^2 - 12x + 27$

55. $t^2 - 9t - 36$

56. $t^2 - 6t - 7$

For the following exercises, factor the polynomial.

57. $7x^2 + 48x - 7$

58. $10h^2 - 9h - 9$

59. $2b^2 + 13b - 24$

60. $9d^2 - 73d + 8$

61. $5t^2 - 19t + 12$

62. $6x^2 + 17x + 7$

Factor the difference of squares.

63. $x^2 - 1$

64. $x^2 - 4$

65. $4m^2 - 9$

66. $16x^2 - 100$

67. $25x^2 - 196$

68. $121p^2 - 169$

69. $144a^2 - 49b^2$

70. $36x^2 - 81y^2$

For the following exercises, solve the quadratic equation by factoring.

71. $x^2 + 8x = 0$

72. $x^2 - 11x = 0$

73. $2a^2 - 14a = 0$

74. $3a^2 + 18a = 0$

75. $x^2 - 121 = 0$

76. $y^2 - 225 = 0$

77. $m^2 - 20m + 100 = 0$

78. $q^2 - 12q + 36 = 0$

79. $4x^2 - 25 = 0$

80. $9x^2 - 16 = 0$

81. $x^2 - 9x + 18 = 0$

82. $x^2 - 11x + 30 = 0$

83. $x^2 + 2x - 35 = 0$

84. $x^2 + 3x - 54 = 0$

85. $9x^2 + 30x + 16 = 0$

86. $8x^2 + 6x - 9 = 0$

87. $2x^2 + 9x - 5 = 0$

88. $6x^2 + 17x + 5 = 0$

For the following exercises, solve the equation by factoring. You will need to set the equation equal to 0 first, then find the greatest common factor, and finally see if further factoring is possible.

89. $4x^2 = 5x$

90. $2x^2 + 14x = 36$

91. $5x^2 = 5x + 30$

92. $4x^2 + 8 = 12x$

93. $3x^2 - 36 = 12x$

94. $2x^2 = 11x$

95. $3x^2 = 75$

96. $6x^2 = 54$

97. $x^3 + 5x^2 = -4x$

98. $x^3 + 7x^2 = -10x$

99. $2x^4 = -30x^3 - 88x^2$

100. $3x^4 = 27x^3 - 60x^2$

Practice Set F — Answers

29. $(x - 6)(x + 1) = 0$; $x = 6, x = -1$

30. $(x - 7)(x + 3) = 0$; $x = 7, x = -3$

31. $(x + 5)(x - 5) = 0$; $x = -5, x = 5$

32. $x(x + 10)(x + 1) = 0$; $x = 0, x = -10, x = -1$

4.2 Quadratic Functions in Standard Form

OVERVIEW

Curved, dish-like antennas are commonly used to focus microwaves and radio waves to transmit television and telephone signals, as well as satellite and spacecraft communication. The cross-section of the antenna is in the shape of a parabola, which can be described by a quadratic function.

In this section, we will investigate quadratic functions, which frequently model problems involving area and projectile motion. You will learn to:

- Know the definition of a quadratic function in standard form.

- Identify characteristics of parabolas.

- Find the vertex and y-intercept of a parabola.

- Determine a quadratic function's domain and range.

- Solve problems involving a quadratic function's minimum or maximum value.

A. QUADRATIC FUNCTIONS AND THEIR GRAPHS

In Section 4.1, we studied how to solve quadratic equations which are equations containing a polynomial in one variable where the highest power on the variable is 2. If we take the polynomial from a quadratic equation and use it as the formula for a function, the result is called a **quadratic function**. We will study quadratic functions in **standard form** as defined below.

Here are some examples of quadratic functions in standard form:

$$f(x) = -4x^2 + 3x + 9, \qquad \text{with constants } a = -4, b = 3, \text{ and } c = 9$$

$$g(x) = x^2 - 5x \qquad \text{with constants } a = 1, b = -5, \text{ and } c = 0$$

$$h(x) = 2.71x^2 - 3.04 \qquad \text{with constants } a = 2.71, b = 0, \text{ and } c = -3.04$$

The ax^2 term is called the **quadratic term**. The bx term is called the **linear term**. The c term, which contains no variable, is simply called the **constant**.

Notice that the constants b and c can equal 0, but the constant a cannot equal 0. If a were allowed to be 0, the ax^2 term would become 0, and the standard form would become $f(x) = bx + c$, which is a linear function, not a quadratic function.

The simplest quadratic function is given by $f(x) = x^2$, and putting that into words, we say that this is a rule that squares the

> ### Quadratic Function
>
> A function that can be written in the form
>
> $$f(x) = ax^2 + bx + c$$
>
> Where a, b, and c are constants with $a \neq 0$, is called a **quadratic function**. We refer to this form as **standard form**.

input to produce the output. In Figure 1, we represent this function numerically with a table of selected input-output pairs. In the Figure 1 table, notice the repetition of most output values. For example, the number 9 shows up twice as an output because when squared, the inputs 3 and –3 both result in the output 9. In general, any input and its opposite will have the same squared output value. This is a form of symmetry we will find useful in our study of quadratic functions.

When we plot the values in the table as ordered pairs and connect them with a smooth curve, we obtain the graph of $f(x) = x^2$ in Figure 2. The curve is called a **parabola**, and every quadratic function $y = ax^2 + bx + c$ has a parabola as the shape of its graph.

In our example, $f(x) = x^2$, the U-shape of the parabola is said to "open upward." As we will see, some parabolas open downward.

x	$y = x^2$
–3	9
–2	4
–1	1
0	0
1	1
2	4
3	9

Figure 1.

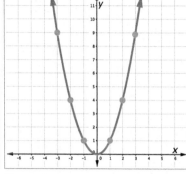

Figure 2.

One important feature of the graph is that it has a turning point, called the **vertex**. For the function, $f(x) = x^2$, the vertex is at $(0, 0)$. The y-coordinate of the vertex, 0, is the minimum output value of the function.

Reading the graph from left to right, we see the curve is decreasing until it gets to the vertex and then increasing after the vertex. Another characteristic of the graph is that it is symmetric with respect to the y-axis. In other words, if we folded the graph along the y-axis, both halves of the graph would lie exactly on top of each other. This symmetry, like the symmetry in the table of values, is because $x^2 = (-x)^2$. A vertical line that intersects a parabola at its vertex is called the **axis of symmetry** or **line of symmetry**.

We see from our Figure 2 graph that the y-intercept of $y = f(x) = x^2$ is the point $(0, 0)$, the point at which the graph crosses the y-axis. Because the x-coordinate of any y-intercept is 0, when we have an equation defining a function, we can find the y-coordinate of the y-intercept by evaluating the function at $x = 0$ — in other words, by finding $f(0)$. For $f(x) = x^2$, $f(0) = 0^2 = 0$, which confirms our graphical finding.

For more complex quadratic functions, we could always find the y-intercept by evaluating $f(0)$. Notice what we get when we use the standard form of the quadratic function, $f(x) = ax^2 + bx + c$, and then evaluate it at $x = 0$. We obtain $f(0) = a(0)^2 + b(0) + c = c$. This is significant. The y-intercept for any quadratic function in standard form is $(0, c)$.

Let's look at another quadratic function, $g(x) = -x^2$. This function differs from $f(x) = x^2$ by the value of the coefficient a. For this new function, $a = -1$, so each input will be squared and then multiplied by –1. Figure 3 shows some input-output pairs for g.

From the graph in Figure 4, we observe that the vertex and y-intercept are at $(0, 0)$ and that the parabola opens downward.

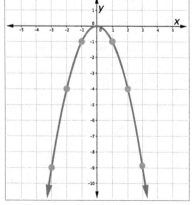

x	$g(x) = -x^2$
–3	–9
–2	–4
–1	–1
0	–0
1	–1
2	–4
3	–9

Figure 3.

Figure 4.

Next, we look at the graphs of a few more quadratic functions to illustrate the effect of the sign of the coefficient a on the orientation of the parabola. See Figure 5.

For $f(x) = -2x^2 + 4x + 5,$ $a = -2$, and the parabola opens down.

For $g(x) = 3x^2 - 12x + 1,$ $a = 3$, and the parabola opens up.

For $h(x) = -\frac{1}{2}x^2 + 3x + 7,$ $a = -\frac{1}{2}$, and the parabola opens down.

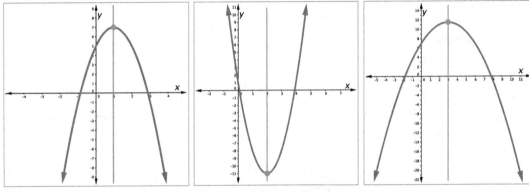

Figure 5. Graphs of $f(x) = -2x^2 + 4x + 5$, $g(x) = 3x^2 - 12x + 1$ and $h(x) = -\frac{1}{2}x^2 + 3x + 7$.

In general, if the coefficient a is positive, the parabola opens up; and if the coefficient a is negative, the parabola opens down. We also observe the graph of each quadratic function in Figure 5 has an axis of symmetry that passes through its vertex. The y-intercept for each graph is at $(0, c)$, where c is the value of the constant.

Our observations about the graphs of quadratic functions can be generalized as follows. A graph of a quadratic function in standard form $f(x) = ax^2 + bx + c$, with $a \neq 0$, has the following characteristics:

◆ The shape of the graph is a parabola.

◆ If $a > 0$, then the parabola opens up. If $a < 0$, then the parabola opens down.

◆ The parabola has a turning point called the vertex.

◆ The parabola is symmetric about the axis of symmetry, a vertical line that intersects the vertex.

◆ The y-intercept is at $(0, c)$.

▶ *Example 1*

Identifying the Characteristics of a Parabola

For the function $f(x) = -2x^2 + 8x + 1$, do the following:

a. Does the parabola open up or down?

b. Make a table of selected values using $x = -1, 0, 1, 2, 3, 4, 5$.

c. Graph the curve.

d. Identify the vertex.

e. Identify the line of symmetry and sketch it as a dashed line on the graph.

f. Identify the y-intercept.

Solution

a. $a = -2$, so $a < 0$ and the parabola opens down.

b. Evaluate the function for the given x-values. For example:

$$f(-1) = -2(-1)^2 + 8(-1) + 1 = -9$$
$$f(0) = -2(0)^2 + 8(0) + 1 = 1$$
$$f(1) = -2(1)^2 + 8(1) + 1 = 7$$

and so on. See Figure 6.

x	f (x)
−1	−9
0	1
1	7
2	9
3	7
4	1
5	−9

Figure 6.

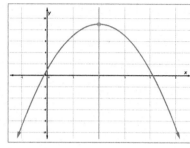

Figure 7.

c. Plot the points in Figure 6 and connect them with a smooth curve in Figure 7.

d. From both the table and the graph, we can see the vertex is at (2, 9).

e. The line of symmetry is the vertical line that intercepts the vertex. See Figure 7.

f. The y-intercept of the graph is at (0, 1).

Practice Set A

Try this problem, and then turn the page to check your work.

1. For the function $f(x) = x^2 + 6x + 2$

   a. Does the parabola open up or open down?

   b. Make a table of selected values using $x = -6, -5, -4, -3, -2, -1, 0$.

   c. Graph the curve. Draw the line of symmetry as a dashed line.

   d. Identify the vertex.

   e. Identify the y-intercept.

B. THE VERTEX FORMULA

To find a formula for the x-coordinate of the vertex of a parabola $f(x) = ax^2 + bx + c$, we look at two cases.

If $b = 0$, the equation is of the form $f(x) = ax^2 + c$. The y-intercept is $(0, c)$ and is the vertex of the parabola. So the x-coordinate of the vertex is 0.

If $b \neq 0$, we can find the x-coordinate of the vertex in a few steps using the symmetry of parabolas. We start by finding the x-coordinate of the point on the parabola that is symmetric to the y-intercept. The point with the same y-coordinate as the y-intercept is c.

$$c = ax^2 + bx + c \quad \text{Substitute } c \text{ for } f(x).$$
$$0 = ax^2 + bx \quad \text{Subtract } c \text{ form both sides.}$$
$$0 = x(ax + b) \quad \text{Factor the right-hand side.}$$

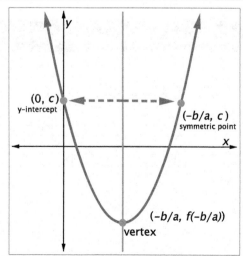

Figure 8.

$x = 0$ or $ax + b = 0$ Apply the zero product property.

$x = 0$ or $ax = -b$ Subtract b from both sides.

$x = 0$ or $x = -\dfrac{b}{a}$ Divide both sides by a.

The x-coordinate of the y-intercept is 0. The x-coordinate of the symmetric point is $-\dfrac{b}{a}$. These points are equidistant from the line of symmetry. See Figure 5. We find the x-coordinate of the vertex by averaging the x-coordinate of the y-intercept and the x-coordinate of the symmetric point:

$x = \dfrac{0 + \left(-\frac{b}{a}\right)}{2}$ Use the definition of average.

$x = \dfrac{\left(-\frac{b}{a}\right)}{2}$ Simplify the numerator.

$x = -\dfrac{b}{a} \cdot \dfrac{1}{2}$ Dividing by 2 is equivalent to multiplying by the reciprocal, $\dfrac{1}{2}$.

$x = -\dfrac{b}{2a}$ Multiply the fractions.

The formula of the x-coordinate of the vertex is thus $x = -\dfrac{b}{2a}$. If $b = 0$, this formula gives $x = -\dfrac{0}{2a} = 0$, which agrees with our earlier statement. Therefore, the formula works for any value of b.

To find the y-coordinate of the vertex, we evaluate the function f for the input $x = -\dfrac{b}{2a}$. That is, we substitute the x-coordinate of the vertex back into the original equation and compute the y-coordinate: $y = f\left(-\dfrac{b}{2a}\right)$

To find the vertex of the graph of a quadratic function $f(x) = ax^2 + bx + c$, do this:

1. Find the x-coordinate of the vertex using the **vertex formula**, $x = -\dfrac{b}{2a}$.

2. Find the y-coordinate of the vertex by evaluating f at the value found in Step 1, $f\left(-\dfrac{b}{2a}\right)$.

Practice Set A — Answers

1.

a. $a = 1$, so the parabola opens up.

b.

x	−6	−5	−4	−3	−2	−1	0
$f(x)$	2	−3	−6	−7	−6	−3	2

c.

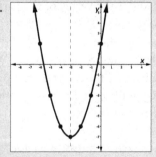

d. vertex: $(-3, -7)$

e. y-intercept: $(0, 2)$

▶ *Example 2*

Using the Vertex Formula to Find the Vertex

   1. $f(x) = -2x^2 + 4x + 5$

   2. $f(x) = x^2 + 5x$

For both of these function,s do the following:

   a. Determine if the parabola opens up or down.

   b. Find the *y*-intercept.

   c. Find the vertex.

   d. Verify your answers using the graph of the function on a graphing calculator.

Solution

1. $f(x) = -2x^2 + 4x + 5$

a. $a = -2$, so the parabola opens down.

b. $c = 5$, so the *y*-intercept is $(0, 5)$.

c. Since $a = -2$ and $b = 4$, the *x*-coordinate of the vertex is: $x = -\dfrac{b}{2a} = -\dfrac{4}{2(-2)} = -\dfrac{4}{-4} = 1$.

   The *y*-coordinate is found by evaluating $f(1)$: $y = f(1) = -2(1)^2 + 4(1) + 5 = 7$.
The vertex is $(1, 7)$.

d. Graph the function in the standard window (ZOOM 6) on the calculator. Make sure any plots are turned off. Use the trace feature on the calculator and enter the *x*-values of the *y*-intercept and the vertex to check your work. See figures 9 and 10.

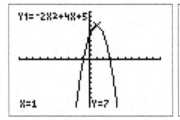

Figure 9

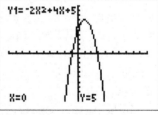

Figure 10.

2. $f(x) = x^2 + 5x$

a. $a = 1$, so the parabola opens up.

b. $c = 0$, so the *y*-intercept is $(0, 0)$.

c. Since $a = 1$ and $b = 5$, the *x*-coordinate of the vertex is: $x = -\dfrac{b}{2a} = -\dfrac{5}{2(1)} = -\dfrac{5}{2} = -2.5$.

   Evaluate $f(-2.5)$ for the *y*-coordinate: $y = f(-2.5) = (-2.5)^2 + 5(-2.5) = -6.25$
The vertex is $(-2.5, -6.25)$.

d. Verify the work as in problem 1 of this example. See Figures 11 and 12.

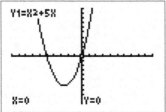

Figure 11

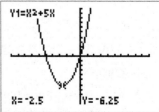

Figure 12.

Practice Set B

Try this problem, and then turn the page to check your work.

2. For the function $f(x) = \frac{1}{2}x^2 - 4x + 3$ **c.** Identify the vertex.

   a. Does the parabola open up or open down?   **d.** Verify your work on a graphing calculator.

   b. Identify the y-intercept .

C. FINDING THE DOMAIN AND RANGE OF A QUADRATIC FUNCTION

Although many of the x-values we've worked with have been integer values, we can let x equal any real number in a quadratic function $f(x) = ax^2 + bx + c$. Therefore, the domain of any quadratic function is all real numbers, unless the context of a real-life application problem presents restrictions.

> ### Domain and Range of a Quadratic Function
>
> For a quadratic function $f(x) = ax^2 + bx + c$ with vertex $\left(-\frac{b}{2a}, f\left(-\frac{b}{2a}\right)\right)$:
> - The domain is all real numbers.
> - If $a > 0$, the parabola opens up and the range is: $y \geq f\left(-\frac{b}{2a}\right)$, or $\left[f\left(-\frac{b}{2a}\right), \infty\right)$.
> - If $a < 0$, the parabola opens down and the range is: $y \leq f\left(-\frac{b}{2a}\right)$, or $\left(-\infty, f\left(-\frac{b}{2a}\right)\right]$.

The range, however, depends on the location of the vertex and whether the parabola opens up or down - things affected by the values of the constants a, b, and c. Because the vertex is a turning point on the graph of the parabola, the range will consist of all y-values greater than or equal to the y-coordinate of the vertex or all y-values less than or equal to the y-coordinate of the vertex, depending on the orientation of the parabola.

If the parabola opens up, the y-coordinate of the vertex is the **minimum value** of the function, as in Figure 15. If the parabola opens down, as in Figure 16, the y-coordinate of the vertex is the **maximum value** of the function.

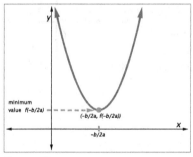

Figure 15

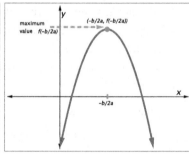

Figure 16

▶ *Example 3*

Finding the Domain and Range of a Quadratic Function

Find the domain and range of $f(x) = -4x^2 - 12x - 3$.

Solution

As with any quadratic function, the domain is all real numbers.

   Because a is negative, the parabola opens downward and the range will be y-values less than or equal to the y-coordinate of the vertex. We begin by finding the x-value of the vertex:

$$x = -\frac{b}{2a} = -\frac{-12}{2(-4)} = -\frac{-12}{-8} = -1.5$$

We find the y-coordinate by evaluating $f(-1.5)$:

$$y = f(-1.5) = -4(-1.5)^2 - 12(-1.5) - 3 = 6$$

The vertex is $(-1.5, 6)$. Since the parabola opens down, the y-coordinate of the vertex is the maximum value of the range. The range is: $(-\infty, 6]$.

▶ Example 4

Finding the Domain and Range of a Quadratic Function

Find the domain and range of the quadratic functions graphed below in Figures 17 to 19.

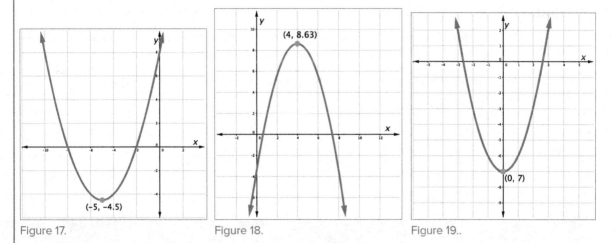

Figure 17.   Figure 18.   Figure 19..

Solution

1. The domain is all real numbers or $(-\infty, \infty)$. The y-coordinate of the vertex is the minimum value of the function. So, the range is $y \geq -4.5$ or we can write $[-4.5, \infty)$.

2. The domain is all real numbers or $(-\infty, \infty)$. The y-coordinate of the vertex is the maximum value of the function. So, the range is $y \leq 8.63$ or we can write $(-\infty, 8.63]$.

3. The domain is all real numbers or $(-\infty, \infty)$. The y-coordinate of the vertex is the minimum value of the function. So, the range is $y \geq -7$ or we can write $[-7, \infty)$.

Practice Set C

Try the following problems. Turn the page to check your work.

3. Find the domain and range of $f(x) = -1.6x^2 - 4.8x + 3$.

4. Find the domain and range of $f(x) = 3x^2 + 12x + 14$.

5. Find the domain and range of a quadratic function with coefficient $a = 1.75$ and vertex at $(5.25, -8.94)$

D. APPLICATIONS USING MAXIMUM OR MINIMUM VALUE

As we've seen, the output of a quadratic function at the vertex is the maximum or minimum value of the function, depending on the orientation of the parabola. There are many real-world scenarios that involve finding the maximum or minimum value of a quadratic function, including applications involving area, projectile motion, and revenue.

▶ *Example 4*

Finding the Maximum Value of a Quadratic Function

A backyard farmer wants to enclose a rectangular space for a new garden within her fenced backyard. She has purchased 80 feet of wire fencing to enclose three sides, and she will use a section of the backyard fence as the fourth side.

   a. Find a formula for the area enclosed by the fence if the sides of fencing perpendicular to the existing fence have length *L*.

   b. What dimensions should she make her garden to maximize the enclosed area?

   c. What is the maximum area?

Practice Set B — Answers

2.

   a. The parabola opens up.

   b. The *y*-intercept is (0, 3).

   c. The vertex is (4, –5).

   d. See figures 13 and 14.

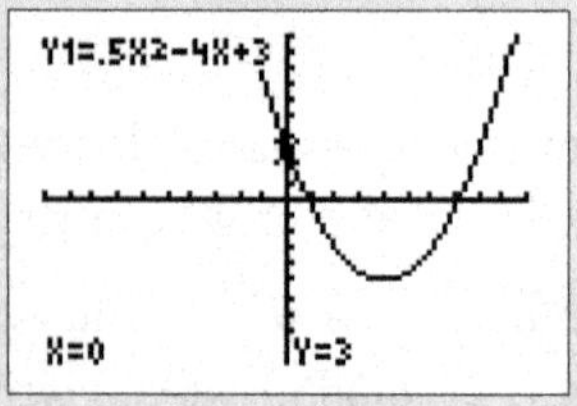

Figure 13.

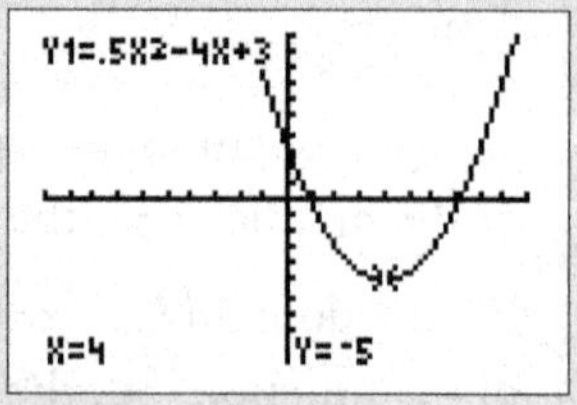

Figure 14.

Practice Set C — Answers

3. The domain is all real numbers. The parabola opens down and the vertex is at (–1.5, 6.6), so the range is $y \le 6.6$, or $(-\infty, 6.6]$.

4. The domain is all real numbers. The parabola opens up and the vertex is at (–2, 2), so the range is $y \ge 2$, or $[2, \infty)$.

5. The domain is all real numbers. The parabola opens up, so the range is $y \ge -8.94$, or $[-8.94, \infty)$.

Solution

We use a diagram such as Figure 20 to record the given information. It is also helpful to introduce a temporary variable, W, to represent the width of the garden and the length of the fence section parallel to the backyard fence.

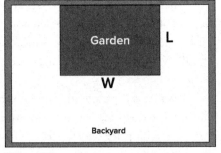

a. We know we have only 80 feet of fence available, and $L + W + L = 80$, or more simply, $2L + W = 80$. This equation allows us to represent the width, W, in terms of L.

Figure 20.

Subtracting $2L$ from both sides we get: $W = 80 - 2L$. Now we are ready to write an equation for the area the fence encloses.

We know the area of a rectangle is length multiplied by width, so:

$$\begin{aligned} A &= LW & \text{Use the area formula for rectangles.} \\ &= L(80{-}2L) & \text{Substitute } W = 80 - 2L. \\ A &= 80L - 2L^2 & \text{Apply the distributive property.} \end{aligned}$$

This formula represents the area of the fence in terms of the variable length L. It is a quadratic function and can be written in standard form:

$$A = f(L) = -2L^2 + 80L$$

b. The quadratic has a negative leading coefficient, $a = -2$, so the parabola will open downward, and the input value (the L-coordinate in this case) of the vertex will give the length that maximizes the area. Using the vertex formula:

$$L = -\frac{b}{2a} = -\frac{80}{2(-2)} = -\frac{80}{-4} = 20$$

So the length is 20 feet and the width is $W = 80 - 2L = 80 - 2(20) = 40$ feet.

The dimensions that will maximize the area of the garden, using 80 feet of fencing are a length (the two shorter sides) of 20 feet and a width of 40 feet.

c. We find the maximum area of the garden by evaluating the function at $L = 20$. This gives the output value of the vertex.

$$A = f(L) = -2(20)^2 + 80(20) = 800$$

The maximum area for the garden is 800 square feet.

This problem can be illustrated by graphing the quadratic function. We can see where the maximum area occurs on a graph of the quadratic function in Figure 21.

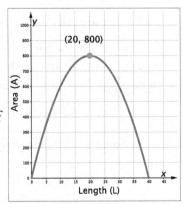

Figure 21.

When a business projects sales and revenue it often takes into account the fact that the unit price of an item affects its supply and demand. That is, if the unit price goes up, the demand for the item will usually decrease. For example, a local newspaper currently has 84,000 subscribers at a quarterly charge of $30. Market research has suggested that if the owners raise the price to $32, they would lose 5,000 subscribers.

Revenue is the amount of money a business brings in. In this case, the revenue can be found by multiplying the price per subscription times the number of subscribers, or quantity. We can introduce variables, p for price per subscription and Q for quantity, giving us the equation Revenue $= pQ$. We use this equation and scenario in the following example.

▶ *Example 5*

Finding Maximum Revenue

A local newspaper estimates that the quantity of subscriptions sold, Q, depends on the quarterly price p in dollars of the subscription according to the equation $Q = -2{,}500p + 159{,}000$.

a. Find a formula for the revenue $R = f(p)$ of the newspaper if Q number of subscriptions is sold at a quarterly price p.

b. Find $f(40)$ and interpret this value in the context of the problem.

c. What price should the newspaper charge in order to maximize revenue?

d. What is the maximum revenue?

Solution

a. The revenue R is found using $R = pQ$, so in this case, that means $R = p(-2{,}500p + 159{,}000)$. Applying the distributive property, we obtain a quadratic function in standard form that models the revenue of the newspaper as a function of the subscription price:

$$R = f(p) = -2500p^2 + 159{,}000p$$

b. To find $f(40)$, substitute $p = 40$ in the equation:

$$f(40) = -2500(40)^2 + 159{,}000(40) = 2{,}360{,}000$$

When the subscription price is $40, the newspaper's revenue is $2,360,000.

c. The leading coefficient of the model is negative so the vertex will provide a maximum value. To find the quarterly subscription price that will maximize revenue for the newspaper, we find the input value (the p-coordinate in this case) of the vertex:

$$p = -\frac{b}{2a} = -\frac{159{,}000}{2(-2500)} = -\frac{159{,}000}{-5000} = 31.8$$

The model tells us that the maximum revenue will occur if the newspaper charges $31.80 for a quarterly subscription.

d. To find the maximum revenue, we evaluate the revenue function at $p = 31.8$:

$$f(31.8) = -2500(31.8)^2 + 159,000(31.8) = 2,528,100$$

The maximum revenue for the newspaper will be $2,528,100 if it charges $31.80 per subscription. We can illustrate the situation by graphing the quadratic function in Figure 22. We see the maximum revenue on the graph.

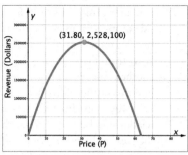

Figure 22.

Exercises

For the quadratic functions in exercises 1 – 4, complete the following:

   a.   Determine if the parabola opens up or down.

   b.   Make a table of selected values using $x = -3, -2, -1, 0, 1, 2, 3, 4$.

   c.   Graph the curve.

   d.   Identify the vertex.

   e.   Identify the y-intercept.

1. $f(x) = 3x^2 + 6x - 5$ **3.** $f(x) = -x^2 + 4x + 3$

2. $f(x) = 2x^2 - 8x + 1$ **4.** $f(x) = -2x^2 + 4x + 6$

For the quadratic functions in exercises 5 – 12 complete the following and verify your answers with a graphing calculator.

   a.   Determine if the parabola opens up or down.

   b.   Find the y-intercept.

   c.   Find the vertex.

5. $f(x) = x^2 - 4$ **9.** $f(x) = -x^2 + 2x + 5$

6. $f(x) = 9 - x^2$ **10.** $f(x) = x^2 - 8x + 16$

7. $f(x) = x^2 + 8x + 13$ **11.** $f(x) = x^2 - x + \dfrac{5}{4}$

8. $f(x) = x^2 - 12x + 39$ **12.** $f(x) = -\dfrac{1}{2}x^2 + 3x$

For exercises 13 – 20, find the vertex of the parabola and determine the domain and range of the function.

13. $f(x) = -4x^2 - 12x - 2$ **19.**

14. $f(x) = -6x^2 - 24x - 25$

15. $f(x) = -2.4x^2 - 4.8x + 3.2$

16. $f(x) = x^2 + 5x + 9$

17. $f(x) = 2.2x^2 - 13.2x + 9.9$

18. $f(x) = -2x^2 + 10x + 7.5$

20.

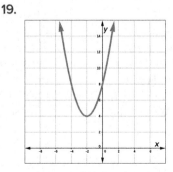

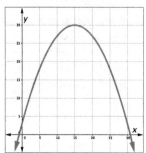

21. A farmer plans to use 130 feet of fencing and a side of his barn to enclose a rectangular pen for two "bummer lambs," which are lambs raised apart from their mothers for one reason or another. The length of the pen L can be written in terms of the width W as follows: $L = 130 - 2W$. The following equation shows the relationship between the width of the pen and the area of the pen A, where the width is in feet and the area is in square feet. $A = W(130 - 2W)$.

    a. Write the equation for the area function in standard quadratic form.

    b. Find the width that will produce the maximum area of the pen.

    c. What is the maximum area?

22. A rancher plans to use 300 feet of fencing to enclose two adjacent rectangular corrals. The length of each corral L can be written in terms of the width W of each corral as follows: $L = 100 - \frac{4}{3}W$. The following equation shows the relationship between the width of each corral and the total area of both corrals A, where the width is in feet and the area is in square feet. $A = W(100 - \frac{4}{3}W)$

    a. Write the equation for the area function in standard quadratic form. Find the width that will produce the maximum area of the corrals.

    b. What is the maximum area?

23. A manufacturer of garden hoses has daily production costs of $C = = 800 - 10x + 0.25x^2$, where C is the total cost (in dollars) and x is the number of hoses produced. How many garden hoses should be produced each day to yield a minimum production cost? What is the minimum production cost?

24. A manufacturer of sprinkler heads has daily production costs of $C = 745 - 12x + 0.12x^2$, where C is the total cost (in dollars) and x is the number of sprinkler heads produced. How many sprinkler heads should be produced each day to yield a minimum production cost? What is the minimum production cost?

25. The height of a ball y (in feet) of a ball thrown by a child is given by $y = -\frac{1}{12}x^2 + 2x + 4$, where x is the horizontal distance (in feet) from where the ball is thrown. What is the maximum height of the ball?

26. The path of a diver off diving platform is given by $y = -\frac{4}{9}x^2 + \frac{24}{9}x + 12$, where y is the height (in feet) and x is the horizontal distance from the end of the diving board. What is the maximum height of the diver?

27. A company's daily profit, P, in dollars, is given by $P = -2x^2 + 120x - 800$, where x is the number of units produced per day. Find x so that the daily profit is a maximum. What is the maximum profit?

28. A service industry has a daily profit, P, in dollars, given by $P = -3.5x^2 + 329x - 700$, where x is the number of people served per day. Find x so that the daily profit is a maximum. What is the maximum profit?

29. Suppose it is known that if 65 apple trees are planted in a certain orchard, the average yield per tree will be 1500 apples per year. For each additional tree planted in the same orchard, the annual yield per tree drops by 20 apples. The number of apples produced per year, P, is dependent on n, the number of additional apple trees planted, and is given by $P = (65 + n)(1500 - 20n)$. Write the equation in standard form and use it to find how many more trees (above 65) should be planted to produce the maximum crop of apples. What is the maximum crop size?

30. It is estimated that an average of 14,000 people will attend a basketball game when the Portland Trailblazers play at the Moda Center, and the admission price is $100. For each $5 added to the price, the attendance will decrease by 280. The revenue per game R is dependent on n the number of $5 increases in price and is given by $R = (100 + 5n)(14,000 - 280n)$ Write the equation in standard form and use it to find the admission price that will produce the largest revenue per game. What is the largest revenue per game?

4.3 The Square Root Property

OVERVIEW

In this section, we will study how to simplify expressions that have square roots. This skill allows us to solve certain quadratic equations using the **square root property**. We have already solved quadratic equations by factoring and applying the **zero product property**, but not all quadratic equations can be solved this way. It is also true that some quadratic equations do not have real number solutions and we will introduce imaginary numbers in order to study these cases.

In this section, you will:

♦ Learn the product property and quotient property for square roots.

♦ Simplify expressions with square roots.

♦ Rationalize the denominator of a radical expression.

♦ Use the square root property to solve quadratic equations.

♦ Learn the definition of an imaginary number.

♦ Solve equations with imaginary and complex solutions.

A. EVALUATING SQUARE ROOTS

When the square root of a number is squared, the result is the original number. Since $4^2 = 16$, the square root of 16 is 4. The square root operation "undoes" squaring just as subtraction "undoes" addition. When a quadratic equation has no linear term (the bx term is absent because $b = 0$), then the only place x appears is in ax^2 and we can undo the squaring by square rooting. However, before we use square rooting in solving quadratic equations, we need a solid understanding of the notation and rules for working with square roots.

In general terms, if k is a positive real number, then the square root of k is a number that, when multiplied by itself, gives k. We write:

If $k > 0$ and $\sqrt{k} = x$, then $x^2 = k$.

We must make an important distinction in this relationship between the two equations above. The equation $x^2 = k$ has *two* solutions because the square root, x, could be positive or negative. This is because multiplying two positive or two negative numbers gives a positive number and the condition $k > 0$ is met. For example, the equation $x^2 = 25$ has two solutions, $x = -5$ and $x = 5$, because $(-5)^2 = 25$ and $(5)^2 = 25$.

However, the equation $\sqrt{k} = x$ has only *one* solution, and that solution is defined to be the positive square root. We call this the **principal square root**, the nonnegative number that when multiplied by itself equals k. The square root obtained using a calculator is the principal square root.

> ### Principal Square Root
>
> The **principal square root** of k is the nonnegative number that, when multiplied by itself, equals k. It is written as a **radical expression**, $\sqrt{k}$.

The principal square root of k is written as $\sqrt{k}$. As Figure 1 illustrates, the symbol is called a **radical**, the term under the symbol is called the **radicand**, and the entire expression is called a **radical expression**.

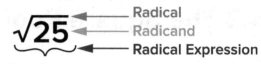

Figure 1.

▶ *Example 1*

Evaluating Square Roots

Evaluate each expression.

   1. $\sqrt{100}$ **2.** $\sqrt{\sqrt{16}}$ **3.** $\sqrt{25 + 144}$ **4.** $\sqrt{49} - \sqrt{81}$

Solution

   1. $\sqrt{100} = 10$ because $10^2 = 100$.

   2. $\sqrt{\sqrt{16}} = \sqrt{4} = 2$ because $4^2 = 16$ and $2^2 = 4$.

   3. $\sqrt{25 + 144} = \sqrt{169} = 13$ because $13^2 = 169$.

   4. $\sqrt{49} - \sqrt{81} = 7 - 9 = -2$ because $7^2 = 49$ and $9^2 = 81$.

For problem 3 in the example above, $\sqrt{25 + 144}$, we *cannot* find the square roots of 25 and 144 seperately before adding. Doing so would result in $\sqrt{25} + \sqrt{144} = 5 + 12 = 17$, which is *not* equivalent to $\sqrt{25 + 144} = \sqrt{169} = 13$. The order of operations requires us to add the terms under the radical symbol before finding the square root. This is because the radical symbol acts as parentheses when the radical bar is extended. In general terms: $\sqrt{a + b} \neq \sqrt{a} + \sqrt{b}$

Practice Set A

Evaluate each expression. When you're finished, turn the page to check your work.

   1. $\sqrt{225}$ **2.** $\sqrt{\sqrt{81}}$ **3.** $\sqrt{25 - 9}$ **4.** $\sqrt{36} + \sqrt{121}$

B. PRODUCT AND QUOTIENT PROPERTIES FOR SQUARE ROOTS

Let's investigate some properties of the square root operation. We have already stated something that is *not* true about square roots, namely $\sqrt{a + b} \neq \sqrt{a} + \sqrt{b}$. In words, the square root of a sum is not equal to the sum of the square roots. This is the case for differences as well: $\sqrt{a - b} \neq \sqrt{a} - \sqrt{b}$. However, there are useful properties of radicals that enable us to simplify radical expressions containing products and quotients.

Using the Product Property to Simplify Square Roots

The first property we will look at is the **product property for square roots**, which allows us to separate the square root of a product of two numbers into the product of two separate radical

The Product Rule for Simplifying Square Roots

If a and b are nonnegative, then

$$\sqrt{ab} = \sqrt{a} \cdot \sqrt{b}$$

   In words, the square root of a product is the product of the square roots.

expressions. For example, we can rewrite $\sqrt{15}$ as $\sqrt{3} \cdot \sqrt{5}$. We can also use the product rule to express the product of multiple radical expressions as a single radical expression.

When asked to simplify a square root, we rewrite it such that there are no perfect squares as factors in the radicand. Remember, **perfect squares** are numbers whose square root is an integer. It is useful to memorize the first 15 perfect squares:

$$1, 4, 9, 16, 25, 36, 49, 64, 81, 100, 121, 144, 169, 196, 225$$

Given a square root expression, we can use the product rule to simplify an expression by following these steps:

1. Factor the largest perfect square from the radicand.

2. Write the radical expression as a product of radical expressions.

3. Simplify.

▶ Example 2

Using the Product Rule to Simplify Square Roots

Simplify the radical expression.

   1. $\sqrt{98}$ **2.** $\sqrt{300}$

Solution

1. The largest perfect square factor of 98 is 49, because 49 • 2 = 98.

$$\begin{aligned} \sqrt{98} &= \sqrt{49 \cdot 2} && \text{Factor the radicand.} \\ &= \sqrt{49} \cdot \sqrt{2} && \text{Apply the product property, } \sqrt{ab} = \sqrt{a} \cdot \sqrt{b}. \\ &= 7\sqrt{2} && \text{Simplify.} \end{aligned}$$

    The expressions $\sqrt{98}$ and $7\sqrt{2}$ are equivalent irrational numbers where the latter is said to be in simplified form. We can check the equivalency on a calculator as shown in Figure 2. The radical expression is said to be in *exact* form. The expression of the number on the calculator in decimal form is *approximate* because the number has been rounded to nine decimal places.

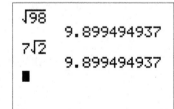

Figure 2. Verify that $\sqrt{98} = 7\sqrt{2}$.

2. The largest perfect square factor of 300 is 100. Note that 25 and 4 are also perfect square factors of 300, just not the largest one.

$$\begin{aligned} \sqrt{300} &= \sqrt{100 \cdot 3} && \text{Factor the radicand.} \\ &= \sqrt{100} \cdot \sqrt{3} && \text{Apply the product property, } \sqrt{ab} = \sqrt{a} \cdot \sqrt{b}. \\ &= 10\sqrt{3} && \text{Simplify.} \end{aligned}$$

If we had mistakenly used 25 as the largest perfect square factor in the first step, we would still arrive at the same simplified radical expression. It would just take a extra few steps.

$$\sqrt{300} = \sqrt{25 \cdot 12} \qquad \text{Factor the radicand using 25 as a factor.}$$

$$= \sqrt{25} \cdot \sqrt{12} \qquad \text{Apply the product property, } \sqrt{ab} = \sqrt{a} \cdot \sqrt{b}.$$

$$= 5\sqrt{12} \qquad \text{Simplify. Notice that } \sqrt{12} \text{ can be factored further.}$$

$$= 5\sqrt{4}\sqrt{3} \qquad \text{Apply the product property again.}$$

$$= 5 \cdot 2\sqrt{3} \qquad \text{Simplify.}$$

$$= 10\sqrt{3} \qquad \text{Multiply.}$$

We can also use the product property to combine radical expressions into one radical expression. Sometimes, when the product is expressed as a single radical, it can be more easily simplified. Other times, it's useful to break a product of two radicals into a product of three or more radicals first. At those times, regrouping the factors may help in the simplification process.

▶ *Example 3*

Using the Product Rule to Simplify the Product of Square Roots

Simplify the radical expression.

1. $\sqrt{12} \cdot \sqrt{3}$

2. $\sqrt{105} \cdot \sqrt{21}$

Solution

1. Use the product rule to combine the radicals.

$$\sqrt{12} \cdot \sqrt{3} = \sqrt{12 \cdot 3} \qquad \text{Express the product as a single radical expression: } \sqrt{ab} = \sqrt{a} \cdot \sqrt{b}.$$

$$= \sqrt{36} \qquad \text{Simplify.}$$

$$= 6$$

2. First factor the radical expressions into the product of smaller radicals.

$$\sqrt{105} \cdot \sqrt{21} = \sqrt{3}\sqrt{5}\sqrt{7} \cdot \sqrt{3}\sqrt{7} \qquad \text{Apply the product property, } \sqrt{ab} = \sqrt{a} \cdot \sqrt{b}.$$

$$= \sqrt{3}\sqrt{3} \cdot \sqrt{7}\sqrt{7} \cdot \sqrt{5} \qquad \text{Reorder the factors.}$$

$$= \sqrt{9} \cdot \sqrt{49} \cdot \sqrt{5} \qquad \text{Express equivalent factors as a single radical expression.}$$

$$= 3 \cdot 7 \cdot \sqrt{5} \qquad \text{Simplify.}$$

$$= 21 \cdot \sqrt{5} \qquad \text{Multiply.}$$

Practice Set A — Answers

1. 15	**2.** 3	**3.** 4	**4.** 17

Using the Quotient Rule to Simplify Square Roots

Just as we can rewrite the square root of a product as a product of square roots, we can also rewrite the square root of a quotient as a quotient of square roots by using the **quotient property for simplifying square roots**. It can be helpful to separate the numerator and denominator of a fraction under a radical so that we can take their square roots separately. We can rewrite $\sqrt{\frac{25}{4}}$ as $\frac{\sqrt{25}}{\sqrt{4}}$. The second fraction is easily simplified to $\frac{5}{2}$.

When given a radical expression, we can use the quotient property of square roots to simplify an expression with these steps:

1. Write the radical expression as the quotient of two radical expressions.

2. Simplify the numerator and denominator.

> **The Quotient Property for Simplifying Square Roots**
>
> If a and b are nonnegative and $b \neq 0$, then
> $$\sqrt{\frac{a}{b}} = \frac{\sqrt{a}}{\sqrt{b}}$$
> In words, the square root of a quotient is the quotient of the square roots.

▶ Example 4

Using the Quotient Rule to Simplify Square Roots

Simplify the radical expression.

1. $\sqrt{\frac{36}{121}}$
2. $\sqrt{\frac{13}{81}}$
3. $\sqrt{\frac{18}{49}}$

Solution

1. $\sqrt{\frac{36}{121}}$

$$\sqrt{\frac{36}{121}} = \frac{\sqrt{36}}{\sqrt{121}} \quad \text{Apply the quotient property, } \sqrt{\frac{a}{b}} = \frac{\sqrt{a}}{\sqrt{b}}.$$
$$= \frac{6}{11} \quad \text{Simplify.}$$

2. $\sqrt{\frac{13}{81}}$

$$\sqrt{\frac{13}{81}} = \frac{\sqrt{13}}{\sqrt{81}} \quad \text{Apply the quotient property, } \sqrt{\frac{a}{b}} = \frac{\sqrt{a}}{\sqrt{b}}.$$
$$= \frac{\sqrt{13}}{9} \quad \text{Simplify the denominator. The radical expression in the numerator can't be simplified.}$$

3. $\sqrt{\frac{18}{49}}$

$$\sqrt{\frac{18}{49}} = \frac{\sqrt{18}}{\sqrt{49}} \quad \text{Apply the quotient property, } \sqrt{\frac{a}{b}} = \frac{\sqrt{a}}{\sqrt{b}}.$$
$$= \frac{\sqrt{18}}{7} \quad \text{Simplify the denominator.}$$
$$= \frac{\sqrt{9}\sqrt{2}}{7} \quad \text{Factor the radicand and apply the product property, } \sqrt{ab} = \sqrt{a} \cdot \sqrt{b}$$
$$= \frac{3\sqrt{2}}{7} \quad \text{Simplify.}$$

Practice Set B

Simplify the radical expressions, and then turn the page to check your work.

5. $\sqrt{150}$ **7.** $\sqrt{3} \cdot \sqrt{15}$ **9.** $\sqrt{\dfrac{63}{64}}$

6. $\sqrt{32}$ **8.** $\sqrt{\dfrac{16}{225}}$

C. RATIONALIZING THE DENOMINATOR OF A RADICAL EXPRESSION

When a radical expression is written in simplest form, mathematical convention says that it will not contain a radical in the denominator. We can remove radicals from the denominators of fractions using a process called **rationalizing the denominator**.

We know that multiplying by 1 does not change the value of an expression. Multiplying by $\frac{3}{3}$ or $\frac{\sqrt{2}}{\sqrt{2}}$ is equivalent to multiplying by 1, so we use this property of multiplication to rewrite expressions that contain radicals in the denominator. To remove radicals from the denominators of fractions, then, multiply by the form of 1 that will eliminate the radical.

For example, to rationalize the denominator of the expression $\frac{5}{\sqrt{2}}$, we multiply the fraction by $\frac{\sqrt{2}}{\sqrt{2}}$ and simplify the denominator:

$$\frac{5}{\sqrt{2}} \cdot \frac{\sqrt{2}}{\sqrt{2}} \;=\; \frac{5\sqrt{2}}{\sqrt{4}} \;=\; \frac{5\sqrt{2}}{2}$$

The last expression in the line above is said to have a rationalized denominator because 2 is a rational number, while $\sqrt{2}$ is an irrational number. We can verify that the original expression is equivalent to the last expression by using a calculator, as in Figure 3.

So when you have an expression with a single square root radical term in the denominator, you can rationalize the denominator by following these steps:

1. Multiply the numerator and denominator by the radical in the denominator.

2. Simplify.

```
5/√2
            3.535533906
5*√2/2
            3.535533906
```

Figure 3 Verifying $\frac{5}{\sqrt{2}} = \frac{5\sqrt{2}}{2}$.

▶ *Example 5*

Rationalizing a Denominator Containing a Single Term

Write the expression in simplest form.

1. $\dfrac{3}{2\sqrt{7}}$ **2.** $\dfrac{2\sqrt{3}}{3\sqrt{10}}$ **3.** $\sqrt{\dfrac{169}{500}}$

Solution

1. The radical in the denominator is $\sqrt{7}$, so multiply the fraction by $\dfrac{\sqrt{7}}{\sqrt{7}}$.

$$\frac{3}{2\sqrt{7}} = \frac{3}{2\sqrt{7}} \bullet \frac{\sqrt{7}}{\sqrt{7}} \qquad \frac{\sqrt{7}}{\sqrt{7}} = 1, \text{ and } a = a \bullet 1.$$

$$\qquad = \frac{3\sqrt{7}}{2\sqrt{49}} \qquad \text{Multiply fractions. Apply the product property in the denominator,}$$
$$\sqrt{ab} = \sqrt{a} \bullet \sqrt{b}.$$

$$\qquad = \frac{3\sqrt{7}}{2 \bullet 7} \qquad \text{Simplify.}$$

$$\qquad = \frac{3\sqrt{7}}{14} \qquad \text{Multiply.}$$

2. The radical in the denominator is $\sqrt{10}$. So multiply the fraction by $\dfrac{\sqrt{10}}{\sqrt{10}}$.

$$\frac{2\sqrt{3}}{3\sqrt{10}} = \frac{2\sqrt{3}}{3\sqrt{10}} \cdot \frac{\sqrt{10}}{\sqrt{10}} \qquad \frac{\sqrt{10}}{\sqrt{10}} = 1, \text{ and } a = a \cdot 1.$$

$$\qquad = \frac{2\sqrt{30}}{3\sqrt{100}} \qquad \text{Multiply fractions. Apply the product property in the denominator:}$$
$$\sqrt{ab} = \sqrt{a} \cdot \sqrt{b}.$$

$$\qquad = \frac{2\sqrt{30}}{30} \qquad \text{Simplify the denominator: } 3\sqrt{100} = 3 \cdot 10 = 30.$$

$$\qquad = \frac{\sqrt{30}}{15} \qquad \text{Simplify the fraction: } \frac{2}{30} = \frac{1}{15}.$$

3. Apply the quotient rule for exponents and simplify. Then rationalize the denominator.

$$\sqrt{\frac{169}{500}} = \frac{\sqrt{169}}{\sqrt{500}} \qquad \text{Apply the quotient property, } \sqrt{\frac{a}{b}} = \frac{\sqrt{a}}{\sqrt{b}}.$$

$$\qquad = \frac{13}{\sqrt{100}\sqrt{5}} \qquad \text{Simplify the numerator. Apply the product property in the denominator:}$$
$$\sqrt{ab} = \sqrt{a} \cdot \sqrt{b}.$$

$$\qquad = \frac{13}{10\sqrt{5}} \qquad \text{Simplify the denominator.}$$

$$\qquad = \frac{13}{10\sqrt{5}} \cdot \frac{\sqrt{5}}{\sqrt{5}} \qquad \text{Multiply by 1. } \frac{\sqrt{5}}{\sqrt{5}} = 1$$

$$\qquad = \frac{13\sqrt{5}}{10\sqrt{25}} \qquad \text{Multiply fractions. Apply the product property in the denominator:}$$
$$\sqrt{ab} = \sqrt{a} \cdot \sqrt{b}.$$

$$\qquad = \frac{13\sqrt{5}}{50} \qquad \text{Simplify the denominator: } 10\sqrt{25} = 10 \cdot 5 = 50.$$

Practice Set C

Write the expression in simplest form and then turn the page to check your work.

10. $\dfrac{4}{\sqrt{3}}$ **11.** $\dfrac{12\sqrt{3}}{\sqrt{2}}$ **12.** $\sqrt{\dfrac{25}{99}}$

D. SOLVING QUADRATIC EQUATIONS USING THE SQUARE ROOT PROPERTY

Now that we have a solid foundation for using and simplifying square roots, we're ready to use square rooting to solve some quadratic equations. To do so we introduce the **square root property,** which gives us an efficient way to solve quadratic equations containing an x^2 term (quadratic) but no x term (linear). With this property, we isolate the x^2 term and take the positive and negative square root of the number on the other side of the equal sign.

> **The Square Root Property**
>
> For $k \geq 0$, the square root property states that if $x^2 = k$, then $x = \pm\sqrt{k}$.

For example if we were to solve $x^2 = 25$ mentally, we know there are two answers, 5 and –5. The square root property gives us the answer as $x = \pm\sqrt{25}$, which simplifies to 5 and –5. Note that since $\sqrt{25}$ by itself is just the principal square root, 5, the ± sign is necessary to pick up both square roots.

When solving a quadratic equation that contains no x term, we can use the square root property by following these steps:

1. Isolate x^2 on the left side of the equal sign.

2. Replace the Step 1 equation with $x =$ the positive or negative square root of the expression on the right side.

3. Simplify the radical expression on the right side, either keeping the ± to represent the two answers or writing the two answers out separately.

The most common student error in applying the square root property is to forget the ± sign. When you forget this symbol, you only obtain one of two possible solutions.

▶ *Example 6*

Solving a Simple Quadratic Equation Using the Square Root Property

Use the square root property to solve the quadratic equation: $x^2 = 8$.

Solution

Apply the square root property, and simplify the radical. Remember to use the ± sign.

$$x^2 = 8 \qquad \text{Original equation.}$$
$$x = \pm\sqrt{8} \qquad \text{Apply the square root property.}$$
$$x = \pm\sqrt{4}\sqrt{2} \qquad \text{Apply the product property.}$$
$$x = \pm 2\sqrt{2} \qquad \text{Simplify.}$$

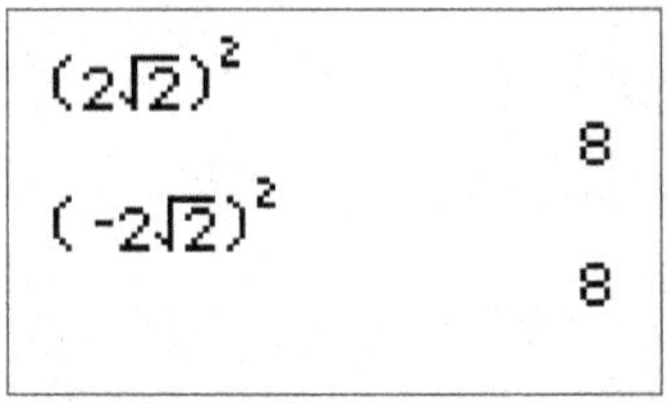

The solutions are $x = 2\sqrt{2}$ and $x = -2\sqrt{2}$. In Figure 4, we use a calculator to verify our work.

$$\left(2\sqrt{2}\right)^2$$
$$ 8$$
$$\left(-2\sqrt{2}\right)^2$$
$$ 8$$

Figure 4. Verify the solutions.

Practice Set B — Answers

5. $5\sqrt{6}$	**6.** $4\sqrt{2}$	**7.** $3\sqrt{5}$	**8.** $\dfrac{4}{15}$	**9.** $\dfrac{3\sqrt{7}}{8}$

▶ *Example 7*

Solving a Quadratic Equation Using the Square Root Property

Solve the quadratic equation: $4x^2 + 1 = 8$.

Solution

First, we isolate x^2. Then we apply the square root property.

$4x^2 + 1 = 8$	Original equation.
$4x^2 = 7$	Subtract 1 from both sides.
$x^2 = \dfrac{7}{4}$	Divide both sides by 4.
$x = \pm\sqrt{\dfrac{7}{4}}$	Apply the square root property.
$x = \pm\dfrac{\sqrt{7}}{2}$	Simplify using the quotient property.

The solutions are $x = \dfrac{\sqrt{7}}{2}$ and $x = -\dfrac{\sqrt{7}}{2}$. This time, we verify the solutions using the "graph and intersect" method on the graphing calculator. Note that $\dfrac{\sqrt{7}}{2} \approx 1.3229$ and $-\dfrac{\sqrt{7}}{2} \approx -1.3229$. See Figure 5.

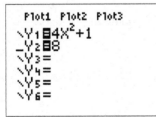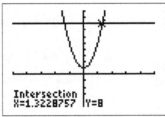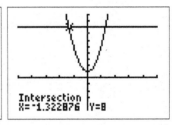

Figure 5. Verify by graph and intersect method.

The square root property allows us to undo the squaring of a variable. But it can also be used to undo the squaring of an expression. We can think of the square root property as $\Box^2 = k$ implies $\Box = \pm\sqrt{k}$ where almost any expression might be in the box. In words, if we have *something* squared equal to a number, then that *something* is equal to plus or minus the square root of the number. So we can use the square root property to solve equations of the form $(ax + b)^2 = k$.

Recall that when solving, we reverse the order of operations for calculations. So we address the expression in parentheses last. This requires us to write the equation with the ± symbol as two separate equations to complete the solving process.

Practice Set C — Answers

10. $\dfrac{4\sqrt{3}}{3}$ **11.** $6\sqrt{6}$ **12.** $\dfrac{5\sqrt{11}}{33}$

▶ *Example 8*

Solve $(5x - 4)^2 = 121$.

Solution

We start by applying the square root property where the squared expression is $5x - 4$.

$$(5x - 4)^2 = 121 \qquad \text{Original equation.}$$
$$5x - 4 = \pm\sqrt{121} \qquad \text{Apply the square root property.}$$
$$5x - 4 = \pm 11 \qquad \text{Simplify the radical expression.}$$
$$5x - 4 = -11 \text{ or } 5x - 4 = 11 \qquad \text{Write as two equations.}$$
$$5x = -7 \text{ or } 5x = 15 \qquad \text{Add 4 to both sides in both equations.}$$
$$x = -\frac{7}{5} \text{ or } x = \frac{15}{5} = 3 \qquad \text{Divide both sides by 5 in both equations.}$$

The two solutions are $x = -\frac{7}{5}$ and $x = 3$.

▶ *Example 9*

Solve $3(x + 7)^2 - 15 = 0$.

Solution

Start by isolating the squared expression, $(x + 7)^2$. Then use the square root property.

$$3(x + 7)^2 - 15 = 0 \qquad \text{Original equation.}$$
$$3(x + 7)^2 = 15 \qquad \text{Add 15 to both sides.}$$
$$(x + 7)^2 = 5 \qquad \text{Divide both sides by 3.}$$
$$x + 7 = \pm\sqrt{5} \qquad \text{Apply the square root property.}$$
$$x + 7 = -\sqrt{5} \text{ or } x + 7 = \sqrt{5} \qquad \text{Write as two equations.}$$
$$x = -7 - \sqrt{5} \text{ or } x = -7 + \sqrt{5} \qquad \text{Subtract 7 from both sides.}$$

When an expression is the combination of a rational and an irrational term, we typically write the rational term first. We can present the two solutions together as $x = -7 \pm \sqrt{5}$.

Practice Set D

Solve for x. When you're finished, turn the page to check your solutions.

13. $x^2 = 20$

14. $7x^2 + 5 = 257$

15. $(4x + 9)^2 = 25$

16. $-10(x - 1)^2 + 270 = 0$

E. COMPLEX NUMBERS AND COMPLEX SOLUTIONS

The Square Root of a Negative Number

We know how to find the square root of any positive real number, so we have been able to solve equations of the form $x^2 = k$, where k is nonnegative. But what if k *is* negative? In section 2.4, we learned that an equation such as $x^2 = -1$ has no real-number solutions. So a new number i is defined as the square root of -1:

$$i = \sqrt{-1}$$

When we square both sides of this identity, we find that $i^2 = -1$:

> ### Square Root of a Negative Number
>
> If k is a positive real number, then $\sqrt{-k} = i\sqrt{k}$.

$$i = \sqrt{-1}$$
$$i^2 = \left(\sqrt{-1}\right)^2$$
$$i^2 = -1$$

We name this new number, the **imaginary unit** i, which is the number whose square is -1. If the value in the radicand is negative, the square root is said to be an imaginary number. We can write the square root of any negative number as a multiple of i. Consider the square root of -49:

$$\sqrt{-49} = i\sqrt{49} = 7i$$

The number $7i$ is of the form bi. A number of the form bi, where b is a real number, is called a **pure imaginary number**.

▶ *Example 10*

Finding the Square Root of a Negative Number

Simplify the expression to bi form, where b is a real number.

1. $\sqrt{-100}$
2. $\sqrt{-12}$
3. $-5\sqrt{-121}$

Solution

1. $\sqrt{-100} = i\sqrt{100} = 10i$
2. $\sqrt{-12} = i\sqrt{12} = i\sqrt{4}\sqrt{3} = 2i\sqrt{3}$
In this case, $b = 2\sqrt{3}$ is a real number. The order of the factors is by convention.
3. $-5\sqrt{-121} = -5i\sqrt{121} = -5i(11) = -55i$

A **complex number** is the sum of a real number and an imaginary number. A complex number is expressed in standard form as $a + bi$, where a is the real part and bi is the imaginary part. For example, $5 + 2i$ is a complex number. So, too, is $3 + 4i\sqrt{3}$. See Figure 6.

$$5 + 2i$$

| Real Part | Imaginary Part |

Figure 6. A complex number in standard form.

Imaginary numbers differ from real numbers in that a squared imaginary number produces a negative real number. Remember that when a *positive* real number is squared, the result is a positive real number. When a *negative* real number is squared, the result is also a positive real number:

$$6^2 = 36 \quad \text{The square of a positive real number is positive.}$$

$$(-6)^2 = 36 \quad \text{The square of a negative real number is positive.}$$

$$(6i)^2 = 6^2 i^2 = 36(-1) = -36 \quad \text{The square of an imaginary number is negative.}$$

Solving Quadratic Equations with Complex Solutions

We can now extend the square root property to include the case where k is negative.

Remember that if we begin with a radical expression then it simplifies to its *one* principal square root. But if we begin with an equation of form $x^2 = k$, then if $k \neq 0$, there will be *two* solutions.

▶ *Example 11*

Solving an Equation with Imaginary Number Solutions

Solve for x.

1. $x^2 = -81$ **2.** $x^2 = -48$ **3.** $6x^2 - 14 = -164$

Solution

1. $x^2 = -81$

$$\begin{aligned} x^2 &= -81 && \text{Start with the original equation.} \\ x &= \pm\sqrt{-81} && \text{Apply the square root property.} \\ x &= \pm i\sqrt{81} && \text{Definition of the square root of a negative number.} \\ x &= \pm 9i && \text{Simplify.} \end{aligned}$$

Practice Set D — Answers

13. $x = \pm 2\sqrt{5}$ **14.** $x = \pm 6$ **15.** $x = -\frac{7}{2}$ and $x = -1$ **16.** $x = 1 \pm 3\sqrt{3}$

2. $x^2 = -48$

$$x^2 = -48$$
$$x = \pm\sqrt{-48} \qquad \text{Apply the square root property.}$$
$$x = \pm i\sqrt{48} \qquad \text{Definition of the square root of a negative number.}$$
$$x = \pm 4i\sqrt{3} \qquad \text{Simplify. } \sqrt{48} = \sqrt{16}\sqrt{3} = 4\sqrt{3}$$

3. $6x^2 - 14 = -164$. We isolate the expression x^2 before applying the square root property.

$$6x^2 - 14 = -164$$
$$6x^2 = -150 \qquad \text{Add 14 to both sides.}$$
$$x^2 = -25 \qquad \text{Divide both sides by 6.}$$
$$x = \pm\sqrt{-25} \qquad \text{Apply the square root property.}$$
$$x = \pm i\sqrt{25} \qquad \text{Definition of the square root of a negative number.}$$
$$x = \pm 5i \qquad \text{Simplify.}$$

If we try to verify our solutions to the last equation using the graph and intersect method on a graphing calculator (Figure 7), we discover that the graphs do not intersect (Figure 7b). This is because the equation has no real-number solutions.

We can verify the solution on the home screen of the calculator (Figure 8). The TI-83 or 84 calculator key pad has the imaginary unit i above the decimal point.

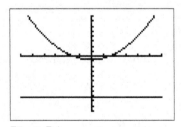

Figure 7.

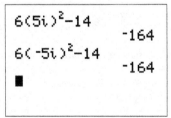

Figure 8. Verify the solutions.

Practice Set E

Simplify the expressions. When you're finished, turn the page to check your work.

17. $\sqrt{-144}$ **18.** $\sqrt{-24}$

Solve the following equations.

19. $x^2 + 10 = -65$ **20.** $2(x-9)^2 + 8 = 0$

▶ *Example 12*

Solving an Equation with Complex Number Solutions

Solve $(x + 5)^2 = -169$.

Solution

$$(x + 5)^2 = -169$$
$$x + 5 = \pm\sqrt{-169} \qquad \text{Apply the square root property.}$$
$$x + 5 = \pm i\sqrt{169} \qquad \text{Definition of the square root of a negative number.}$$
$$x + 5 = \pm 13i \qquad \text{Simplify.}$$
$$x = -5 \pm 13i \qquad \text{Write the solutions in standard complex form.}$$

Exercises

Simplify.

1. $\sqrt{196}$

2. $\sqrt{225}$

3. $\sqrt{28}$

4. $\sqrt{54}$

5. $\sqrt{30} \cdot \sqrt{6}$

6. $\sqrt{24} \cdot \sqrt{3}$

7. $\sqrt{\dfrac{16}{81}}$

8. $\sqrt{\dfrac{49}{121}}$

9. $\sqrt{\dfrac{17}{144}}$

10. $\sqrt{\dfrac{7}{25}}$

11. $\dfrac{6}{\sqrt{5}}$

12. $\dfrac{4}{\sqrt{13}}$

13. $\dfrac{3}{\sqrt{63}}$

14. $\dfrac{10}{\sqrt{125}}$

15. $\sqrt{\dfrac{3}{7}}$

16. $\sqrt{\dfrac{11}{20}}$

Solve using the square root property. Solutions are real numbers.

17. $x^2 = 36$

18. $x^2 = 49$

19. $x^2 - 14 = 0$

20. $x^2 - 23 = 0$

21. $p^2 = 40$

22. $t^2 = 288$

23. $9t^2 = 2$

24. $16p^2 = 29$

25. $6n^2 - 17 = 7$

26. $8n^2 + 43 = 243$

27. $(3x - 8)^2 = 49$

28. $(4x + 3)^2 = 121$

29. $(x + 6)^2 = 7$

30. $(x - 13)^2 = 5$

31. $(x + 9)^2 = 12$

32. $(x - 10)^2 = 27$

33. $5(x - 1)^2 + 42 = 447$

34. $4(x + 3)^2 - 19 = 125$

Practice Set E Answers

17. $12i$

18. $2i\sqrt{6}$

19. $x = \pm 5i\sqrt{3}$

20. $x = 9 \pm 2i$

Simplify.

35. $\sqrt{-9}$

36. $\sqrt{-64}$

37. $\sqrt{-50}$

38. $\sqrt{-8}$

39. $\sqrt{-\dfrac{4}{25}}$

40. $\sqrt{-\dfrac{64}{169}}$

41. $2\sqrt{-48}$

42. $3\sqrt{-20}$

Find all complex-number solutions.

43. $x^2 = -4$

44. $x^2 = -49$

45. $x^2 = -45$

46. $x^2 = -18$

47. $8x^2 + 33 = 17$

48. $5x^2 + 41 = -14$

49. $(x - 9)^2 = -100$

50. $(x - 2)^2 = -25$

51. $3(x + 1)^2 + 58 = 34$

52. $3(x + 8)^2 + 261 = 111$

4.4 The Quadratic Formula

OVERVIEW

We continue to study methods for solving quadratic equations. Factoring and using the square root property are each methods that only work for some equations. In this section we use an important equation called the quadratic formula, which can be used to solve *any* quadratic equation. We also study the connection between the solutions of a quadratic equation and the graph of the related quadratic function.

In this section you will:

♦ Solve quadratic equations by using the quadratic formula.

♦ Find *x*-intercepts of quadratic functions using the quadratic formula.

♦ Find *x*-intercepts of quadratic functions using a graphing calculator.

♦ Use the discriminant to determine the number and type of solutions of a quadratic equation.

♦ Make predictions with a quadratic model.

A. THE QUADRATIC FORMULA

A third method of solving quadratic equations is by using the quadratic formula, which will solve all quadratic equations. We can derive the quadratic formula by yet another solving method called "completing the square." Since the method for completing the square is not covered in this course, we'll omit the derivation of the formula and just present the result below.

A quadratic equation can be solved using the quadratic formula by following these steps:

1. Make sure the equation is in standard form: $ax^2 + bx + c = 0$.

2. Make note of the values of the coefficients and constant term, *a*, *b*, and *c*.

3. Carefully substitute the values noted in step 2 into the equation. To avoid careless errors, use parentheses around each number input into the formula.

4. Calculate and solve.

> ### The Quadratic Formula
>
> Written in standard form, $ax^2 + bx + c = 0$, where *a*, *b*, and *c* are real numbers and $a \neq 0$, any quadratic equation can be solved using the **quadratic formula**:
>
> $$x = \frac{-b \pm \sqrt{b^2 - 4ac}}{2a}$$

Although the quadratic formula works on any quadratic equation in standard form, it's easy to make errors in substituting the values into the formula and simplifying the resulting expression. Pay close attention when substituting, and use parentheses, especially when inserting a negative number. You will also need to make sure both terms in the numerator are divided by the denominator.

▶ *Example 1*

Solve the Quadratic Equation Using the Quadratic Formula

For $x^2 + 2x - 15 = 0$,

   a. Solve for x using the quadratic formula.

   b. Graph the related quadratic function, $y = x^2 + 2x - 15$, and note the coordinates of the x-intercepts.

Solution

a. Identify the coefficients: $a = 1$, $b = 2$, $c = -15$. Then use the quadratic formula.

$$x = \frac{-b \pm \sqrt{b^2 - 4ac}}{2a} \qquad \text{The quadratic formula.}$$

$$x = \frac{-(2) \pm \sqrt{(2)^2 - 4(1)(-15)}}{2(1)} \qquad \text{Substitute: } a = 1, b = 2, \text{ and } c = -15.$$

$$x = \frac{-2 \pm \sqrt{64}}{2} \qquad \text{Simplify: } (2)^2 - 4(1)(-15) = 4 - ^-60 = 64.$$

$$x = \frac{-2 \pm 8}{2} \qquad \text{Simplify further: } \sqrt{64} = 8.$$

$$x = \frac{-2 + 8}{2} \quad \text{or} \quad x = \frac{-2 - 8}{2} \qquad \text{Write as two equations.}$$

$$x = 3 \quad \text{or} \quad x = -5 \qquad \text{Compute.}$$

Notice in the second to last step, the fraction bar is extended below both terms in the numerator. *Both* terms are divided by 2. In this problem, the calculation is easy to do in your head, but if we use a calculator to simplify an expression such as this, we must use two separate steps or put parentheses around the numerator. See Figure 1.

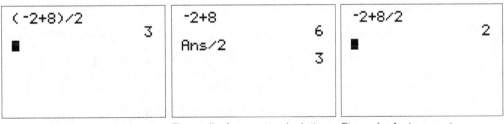

Figure 1a. A correct calculation of $x = \frac{-2+8}{2}$ in one step.

Figure 1b. A correct calculation of $x = \frac{-2+8}{2}$ in two steps.

Figure 1c. An incorrect calculation of $x = \frac{-2+8}{2}$.

b. For $x^2 + 2x - 15 = 0$, the related quadratic function is $f(x) = x^2 + 2x - 15$. Enter the equation into the calculator and find a good viewing window. Enter the solutions to the quadratic equation, $x = 3$ and $x = -5$, as values in $\boxed{\text{TRACE}}$ mode. See Figure 2.

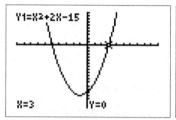

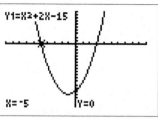

Figure 2a. An x-intercept of the graph.

Figure 2b. An x-intercept of the graph

Note the solutions to the quadratic equation are the x-coordinates of the x-intercepts of the graph of the quadratic function, namely (3, 0) and (–5, 0). In other words, the input values $x = 3$ and $x = -5$ both give an output of $y = 0$.

We can also solve the equation used in Example 1 by factoring and applying the zero product property: $x^2 + 2x - 15 = 0$ becomes $(x - 3)(x + 5) = 0$, so either

$$x - 3 = 0 \quad \text{or} \quad x + 5 = 0 \text{ and}$$
$$x = 3 \quad \text{or} \quad x = -5.$$

We obtain the same solutions using either method. The next example has a quadratic equation that cannot be factored, so we use the quadratic formula to solve.

▶ *Example 2*

Solving a Quadratic Equation with the Quadratic Formula

For $3x^2 = 9x - 4$,

    a. Solve for x using the quadratic formula.

    b. Graph the related quadratic function and verify that the solutions are the x-coordinates of the x-intercepts.

Solution

a. We rewrite the equation in standard form by setting the equation equal to 0.

$$3x^2 = 9x - 4 \qquad \text{Original equation.}$$
$$3x^2 - 9x = -4 \qquad \text{Subtract 9x from both sides.}$$
$$3x^2 - 9x + 4 = 0 \qquad \text{Add 4 to both sides.}$$

Identify the coefficients: $a = 3$, $b = -9$, $c = 4$. Then use the quadratic formula.

$$x = \frac{-(-9) \pm \sqrt{(-9)^2 - 4(3)(4)}}{2(3)} \quad \text{Substitute } a = 3, b = -9, \text{ and } c = 4 \text{ in the quadratic formula.}$$

$$x = \frac{9 \pm \sqrt{33}}{6} \qquad \text{Simplify. } (-9)^2 - 4(3)(4) = 81 - 48 = 33$$

The solutions to the equation are $x = \dfrac{9 + \sqrt{33}}{6}$ and $x = \dfrac{9 - \sqrt{33}}{6}$. Rounded to three decimal places the approximate solutions are: $x = \dfrac{9 + \sqrt{33}}{6} \approx 2.457$ and $x = \dfrac{9 - \sqrt{33}}{6} \approx 0.543$.

b. The related quadratic function is $f(x) = 3x^2 - 9x + 4$. Once again, the solutions to the quadratic equation are the x-coordinates of the x-intercepts of the graph of the parabola, as you see in Figure 3. Notice the parentheses used when entering the x value.

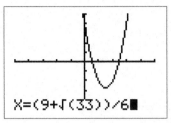 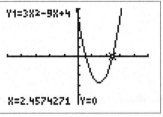 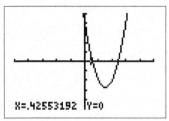

Figure 3a. Entering a radical expression for x.

Figure 3b. An x-intercept of the graph

Figure 3c. An x-intercept of the graph.

▶ Example 3

Solving a Quadratic Equation with the Quadratic Formula

For $4x^2 - 20x + 25 = 0$,

    a. Solve for x using the quadratic formula.

    b. Graph the related quadratic function and verify that the solution is the x-coordinate of the x-intercept.

Solution

a. Identify the coefficients: $a = 4$, $b = -20$, $c = 25$. Then use the quadratic formula.

$$x = \frac{-(-20) \pm \sqrt{(-20)^2 - 4(4)(25)}}{2(4)}$$
    Substitute $a = 4$, $b = -20$, and $c = 25$ in the quadratic formula.

$$x = \frac{20 \pm \sqrt{0}}{8}$$
    Simplify: $(-20)^2 - 4(4)(25) = 400 - 400 = 0$

$$x = \frac{20}{8}$$
    $\sqrt{0} = 0$

$$x = \frac{5}{2} = 2.5$$
    Reduce the fraction.

    This time we found only one solution, $x = 2.5$.

b. The related quadratic function is $f(x) = 4x^2 - 20x + 25$. When we look at the graph of the parabola this time, we see in Figure 4 that it has only one x-intercept at $(2.5, 0)$. We found one solution rather than two because when we evaluated the radical in the quadratic formula we obtained $\pm\sqrt{0} = 0$. Adding or subtracting 0 to a number does not change the number.

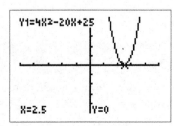

Figure 4.

    The original equation, $4x^2 - 20x + 25 = 0$, can be written in factored form as $(2x - 5)(2x - 5) = (2x - 5)^2 = 0$. Because the factors are identical, we only need to set one of factors equal to zero to find the solution. Solving $2x - 5 = 0$ gives $x = \frac{5}{2}$, which is sometimes referred to as a **double solution**.

In Example 4, we use the quadratic formula to solve a quadratic equation that has imaginary-number solutions.

▶ *Example 4*

Solving a Quadratic Equation with the Quadratic Formula

For $-5x^2 - 5 = -8x$,

- **a.** Solve for x using the quadratic formula.
- **b.** Graph the related quadratic function and note any x- or y-intercepts.

Solution

a. We rewrite the equation in standard form by setting the equation equal to 0.

$$-5x^2 - 5 = -8x \qquad \text{Original equation.}$$

$$-5x^2 + 8x - 5 = 0 \qquad \text{Add } 8x \text{ to both sides and write in standard form: } ax^2 + bx + c = 0.$$

Identify the coefficients: $a = -5$, $b = 8$, $c = -5$. Then use the quadratic formula.

$$x = \frac{-(8) \pm \sqrt{(8)^2 - 4(-5)(-5)}}{2(-5)} \qquad \text{Substitute } a = -5, b = 8, \text{ and } c = -5 \text{ in the quadratic formula.}$$

$$x = \frac{-8 \pm \sqrt{-36}}{-10} \qquad \text{Simplify: } (8)^2 - 4(-5)(-5) = 64 - 100 = -36.$$

$$x = \frac{-8 \pm 6i}{-10} \qquad \text{Square root of a negative number:} \sqrt{-k} = i\sqrt{k}.$$

$$x = \frac{4 \pm 3i}{5} \qquad \text{Reduce the fraction. Divide each term by } -2.$$

In the last step, we divide the expression $\pm 6i$ by -2, which results in $-3i$ and $3i$. These two results could be shortened to $\mp 3i$, but since the order of the two solutions does not matter, we write the more traditional form, $\pm 3i$.

The solutions are two complex numbers: $x = \frac{4 + 3i}{5}$ and $x = \frac{4 - 3i}{5}$. Recall that complex numbers in standard form, $a + bi$, have a real and an imaginary part. We write the solutions in standard complex form by dividing both terms of the numerator by 5: $x = \frac{4}{5} + \frac{3}{5}i$ and $x = \frac{4}{5} - \frac{3}{5}i$.

Practice Set A

Solve the following quadratic equations using the quadratic formula. When you're finished, turn the page to check your work.

1. $9x^2 + 3x - 2 = 0$ **2.** $3x^2 + 2 = 12x$ **3.** $2x^2 - 4x + 6 = 0$

b. Graph the quadratic function, $f(x) = -5x^2 + 8x - 5$. The y-intercept is $(0, -5)$. There are no x-intercepts. See Figure 5. This is related to the fact that we found no real-numbered solutions to the equation $-5x^2 + 8x - 5 = 0$.

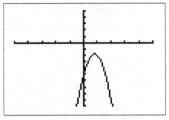

Figure 5. $f(x) = -5x^2 + 8x - 5$

B. FINDING X-INTERCEPTS OF QUADRATIC FUNCTIONS

Examples 1 through 4 demonstrate that the x-intercepts of a quadratic function can be found by setting the output of the function $y = f(x)$ equal to 0 and solving for x. Every x-intercept is a point with y-coordinate equal to 0. Every y-intercept is a point with x-coordinate equal to 0. Recall from Section 4.2 that for a quadratic function in standard form, $f(x) = ax^2 + bx + c$, the y-intercept is at $(0, c)$.

▶ Example 5

Finding the *y*- and *x*-Intercepts of a Parabola

Find the y and x-intercepts of the quadratic function $f(x) = 3x^2 + 5x - 2$. Verify your work by graphing the function on a graphing calculator.

Quadratic Formula Program

You should solve a few quadratic equations by hand using the quadratic formula. Once you understand how to apply the formula, your professor may want you to download a **quadratic formula program** to your calculator. This will allow you to solve more complex or multifaceted problems quickly without getting bogged down in one computation. Ask your professor for more information.

Solution

The function is in standard form with $c = -2$. The y-intercept is at $(0, -2)$. For the x-intercepts, we find all solutions of $f(x) = 0$:

$$0 = 3x^2 + 5x - 2$$

Apply the quadratic formula with $a = 3$, $b = 5$, and $c = -2$:

$$x = \frac{-(5) \pm \sqrt{(5)^2 - 4(3)(-2)}}{2(3)} = \frac{-5 \pm \sqrt{49}}{6} = \frac{-5 \pm 7}{6}$$

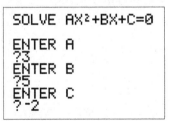

Figure 6.

The last expression simplifies to the solutions $x = \frac{1}{3}$ and $x = -2$. The x-intercepts are $(\frac{1}{3}, 0)$ and $(-2, 0)$.

The equation, $0 = 3x^2 + 5x - 2$, can also be solved using a quadratic formula program on a calculator, as in Figure 6. Notice that the program is set to round solutions to three decimal places.

We verify our work by graphing $f(x)$ on a calculator and entering the values $x = 0$, $x = \frac{1}{3}$, and $x = -2$ while in TRACE mode. Figure 7 shows one of these checks.

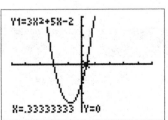

Figure 7. Verify the work.

In Example 6, we estimate the x-intercepts of a quadratic function using the graph-and-intersect method first demonstrated in Section 3.3.

▶ *Example 6*

Finding x-intercepts Using a Graph on a Graphing Calculator

Use a graphing calculator and the graph of the function to estimate the x-intercepts of the quadratic function $f(x) = -0.5x^2 - 3x + 4$. Round values to three decimal places.

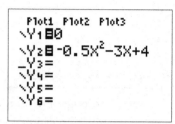

Solution

Since the y-coordinate of the x-intercepts is 0, we are solving this equation:

$$0 = -0.5x^2 - 3x + 4$$

Enter the left side of the equation as Y_1, and enter the right side of the equation as Y_2. Next, use the intersect feature under the [CALC] menu above the [TRACE] button. Figure 8 shows how this works.

Figure 8.

The approximate coordinates of the x-intercepts are (-7.123, 0) and (1.123, 0). Notice in Figure 9 that the y-coordinate is written in scientific notation. Recall from Section 2.1, that $y = 1E-12$, implies $y = 1 \times 10^{-12} = 0.000000000001 \approx 0$. So the y-coordinate of the x-intercept is approximately 0.

When using a calculator to find or verify x-intercepts, the displayed y-value may differ slightly depending on the "guess" entered during the intersection procedure, but the y-value will always round to zero.

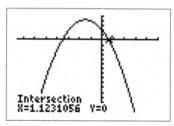
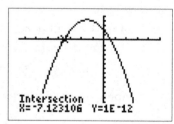

Figure 9.

Practice Set A — Answers

1. $x = -\dfrac{2}{3}, x = \dfrac{1}{3}$

2. $x = \dfrac{6 \pm \sqrt{30}}{3}$

3. $x = \dfrac{4 \pm i\sqrt{32}}{4} = \dfrac{4 \pm 4i\sqrt{2}}{4} = 1 \pm i\sqrt{2}$

In the next example we interpret the intercepts of a graph in the context of a realistic situation.

▶ Example 7

Finding and Interpreting Intercepts of a Quadratic Model

A batter hits a fast pitch at a baseball game. The height of the baseball (in feet) t seconds after it has been hit can be modeled by the function $f(t) = -16t^2 + 120t + 3.5$.

a. Graph the function on a calculator in the window [-1, 10, 1] by [-20, 240, 20].

b. Find and interpret $f(0)$.

c. Find the t-intercepts of the graph of the function and interpret their meaning in the context of the problem.

d. When is the baseball at a height of 80 feet?

Solution

a. Figure 10 shows the graph of the function.

b. $f(0) = -16(0)^2 + 120(0) + 3.5 = 3.5$. The y-intercept of the parabola is (0, 3.5). In the context of this problem, that means that at time $t = 0$, when the batter hits the ball, the height of the baseball is 3.5 feet.

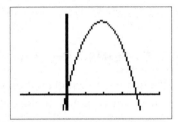

Figure 10.

c. We find the t-intercepts of the parabola by solving the following equation for t:

$$0 = -16t^2 + 120t + 3.5$$

This is most easily solved with a quadratic formula program on a calculator where $a = -16$, $b = 120$, and $c = 3.5$. The solutions, rounded to three decimal places, are $t \approx -0.029$ and $t \approx 7.529$, so the t-intercepts are approximately (–0.029, 0) and (7.529, 0).

The first solution, $t \approx -0.029$, is a negative value that would suggest going back in time before the baseball was hit by the batter. This value makes no sense in the context of the problem and is an example of model breakdown.

The second solution, $t \approx 7.529$, means that the height of the ball will be 0 feet after approximately 7.529 seconds. Or more plainly, it will take 7.529 seconds for the baseball to land.

d. To find when the baseball is at a height of 80, we substitute 80 for $f(t)$ in the equation:

$$80 = -16t^2 + 120t + 3.5$$

We solve this equation using the graph-intersect method on a graphing calculator, where $Y_1 = 80$ and $Y_2 = -16t^2 + 120t + 3.5$. One solution is $t \approx 0.703$ and is shown in Figure 11. The second solution is $t \approx 6.797$.

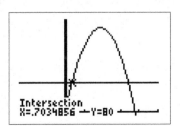

Intersection
X=.7034856   Y=80

Figure 11.

We can also set the equation equal to 0 by subtracting 80 from both sides to obtain standard form:

$$0 = -16t^2 + 120t - 76.5$$

We can quickly solve this equation using a quadratic formula program, and of course, we obtain the same solutions. The baseball will be at a height of 80 at approximately 0.703 seconds on its way up and at 6.797 seconds on its way back down.

Practice Set B

Try the following problems, and then turn the page to check your work.

4. Find the x- and y-intercepts of $f(x) = -2x^2 + 6x - 8$.

5. Use a graphing calculator to estimate the x-intercepts of $f(x) = -1.8x^2 + 13.2x + 24.9$. Round values to three decimal places.

6. A small rocket is launched so that the height of the rocket (in meters) t seconds after it has been launched can be modeled by the function $f(t) = -4.9t^2 + 51t + 2$. Find and interpret the t-intercepts in the context of the problem.

C. DETERMINING THE NUMBER AND TYPE OF SOLUTIONS

The quadratic formula not only generates the solutions to a quadratic equation, it also tells us about the nature of the solutions. To determine the nature of the solutions we need only consider the **discriminant**, which is the expression under the radical, $b^2 - 4ac$:

$$x = \frac{-b \pm \sqrt{b^2 - 4ac}}{2a}$$

If the discriminant is positive, its square root will be a real number and the quadratic formula gives two real-numbered solutions. If the discriminant is negative, its square root is an imaginary number and the quadratic formula gives two imaginary-number solutions. Finally if the discriminant is 0, the quadratic formula gives:

> ### The Discriminant
>
> For $ax^2 + bx + c = 0$, where a, b, and c are real numbers, the **discriminant** is the expression under the radical in the quadratic formula: $b^2 - 4ac$. It tells us whether the solutions are real numbers or imaginary numbers and how many solutions of each type to expect.

$$x = \frac{-b \pm \sqrt{0}}{2a} = \frac{-b}{2a}$$

Therefore, there is one real-number solution. The table below relates the value of the discriminant to the solutions of a quadratic equation.

Value of Discriminant	Results
$b^2 - 4ac = 0$	One real-number solution
$b^2 - 4ac > 0$	Two real-number solutions
$b^2 - 4ac < 0$	Two imaginary-number solutions

▶ *Example 8*

Using the Discriminant to Find the Number and Type of Solutions

Use the discriminant to find the number and type of solutions to the following quadratic equations:

1. $x^2 + 4x + 4 = 0$ 2. $5x^2 - 11x - 3 = 0$ 3. $3x^2 - 10x + 15 = 0$

Solution

Calculate the discriminant $b^2 - 4ac$ for each equation and state the expected type of solutions.

1. $b^2 - 4ac = (4)^2 - 4(1)(4) = 0$. There will be one real-number solution.

2. $b^2 - 4ac = (-11)^2 - 4(5)(-3) = 181 > 0$. There will be two real-number solutions.

3. $b^2 - 4ac = (-10)^2 - 4(3)(15) = -80 < 0$. There will be two imaginary-number solutions.

We've seen that the number of x-intercepts of a parabola can vary depending upon the location of the graph. There is a direct correlation between the value of the discriminant of a quadratic equation and the number of x-intercepts of the corresponding quadratic function. The table below presents this correlation.

Discriminant	*Solutions to $ax^2 + bx + c = 0$*	*x-intercepts of $f(x) = ax^2 + bx + c$*
$b^2 - 4ac = 0$	One real-number solution	One x-intercept (the same point as the vertex)
$b^2 - 4ac > 0$	Two real-number solutions	Two x-intercepts (the graph crosses the x-axis twice)
$b^2 - 4ac < 0$	Two imaginary-number solutions	No x-intercepts (the graph never crosses the x-axis)

Figures 12, 13, and 14 provide examples of each case.

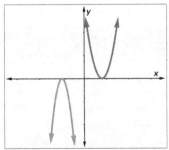 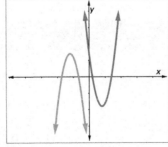 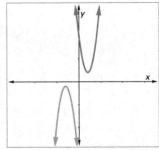

Figure 12. $b^2 - 4ac = 0$, one x-intercept.   Figure 13. $b^2 - 4ac > 0$, two x-intercepts.   Figure 14. $b^2 - 4ac < 0$, no x-intercept.

Practice Set C

For each quadratic equation: find the discriminant; state the number and type of solutions; and state the number of x-intercepts of the corresponding quadratic equation. Turn the page to check your work.

7. $6x^2 + 5x - 2 = 0$ **8.** $0 = -3x^2 + 2x - 8$ **9.** $x^2 - 10x + 25 = 0$

D. DECIDING WHICH SOLVING METHOD TO USE

In this chapter we have discussed three ways to solve quadratic equations: factoring, the square root property, and the quadratic formula. Of course we can also solve an equation graphically with the graph-intersect method. How do we decide which method to use?

Here are some guidelines to help you decide which method to use to solve a quadratic equation:

- Use **factoring** for equations that can be put in the form $ax^2 + bx + c = 0$ and can be easily factored. If $(x - p)(x - q) = 0$, then $x = p$ or $x = q$. Most quadratic equations are prime, so only a small percentage can be solved by factoring. However, if an equation *can* be solved by simple factoring, this is often the easiest method.

- Use **square root property** for equations that can easily be put into the form $x^2 = k$ or $(ax + b)^2 = k$ and then $x = \pm k$ or $ax + b = \pm k$.

- Use **graphing methods** for finding real solutions only. This is most useful when other related questions are asked that can be solved with the graphing features of the calculator.

- Use the **quadratic formula** for any quadratic equation — especially those that cannot be easily solved by another method.

▶ *Example 9*

Deciding Which Solving Method to Use

Solve.

1. $2x^2 - 29 = 5$

2. $x^2 - 8x + 12 = 0$

3. $(x + 6)(x - 7) = 3(x + 2)$

Practice Set B — Answers

4. The y-intercept is $(0, -8)$. The x-intercepts are $(-4, 0)$ and $(1, 0)$.

5. $(-1.556, 0)$ and $(8.889, 0)$

6. The approximate values for the t-intercepts are $(-0.039, 0)$ and $(10.447, 0)$. The first point has a negative value and makes no sense in the context of the problem (model breakdown). The second point is interpreted to mean the rocket will land after approximately 8.889 seconds.

Practice Set C — Answers

7. discriminant = 73; two real-number solutions; two x-intercepts

8. discriminant = -92; two imaginary-number solutions; no x-intercepts

9. discriminant = 0; one real-number solution; one x-intercept

Solution

1. The equation can easily be put into the form $x^2 = k$.

$2x^2 - 29 = 5$	Original equation.
$2x^2 = 34$	Add 29 to both sides.
$x^2 = 17$	Divide both sides by 2.
$x = \pm\sqrt{17}$	Apply the square root property.

2. The equation is easily factorable, so $x^2 - 8x + 12 = 0$ becomes $(x - 6)(x - 8) = 0$. Therefore, $x - 6 = 0$ or $x - 8 = 0$ and $x = 6$ or $x = 8$.

3. First we write the equation in $ax^2 + bx + c = 0$ form:

$(x + 6)(x - 7) = 3(x + 2)$	Original equation.
$x^2 - x - 42 = 3x + 6$	Apply the distributive property.
$x^2 - 4x - 48 = 0$	Set equal to 0 by subtracting $3x$ and 6 from both sides.

The trinomial $x^2 - 4x - 48$ is prime, so we cannot solve by factoring. Since there is a quadratic term and a linear term, it can't be put into the form $x^2 = k$ or easily put into the form $(ax + b)^2 = k$. We must use the quadratic formula with $a = 1$, $b = -4$, and $c = -48$.

The exact solutions are $x = \dfrac{4 \pm \sqrt{208}}{2} = \dfrac{4 \pm 4\sqrt{13}}{2} = 2 \pm 2\sqrt{13}$ Using a quadratic formula program we find approximate solutions: $x \approx 9.211$ and $x \approx -5.211$.

You will have the opportunity to use all of the methods we have discussed for solving quadratic equations as you work the exercises.

Exercises

Use the quadratic formula to solve the equation for all real and non-real solutions. Write the solutions in exact form.

1. $-2x^2 + 7x + 4 = 0$

2. $3x^2 + 5x - 8 = 0$

3. $5x^2 - 6x = 3$

4. $x^2 - 9x = -13$

5. $6x^2 = 2x - 2$

6. $4x^2 = 28x - 49$

7. $9x^2 + 24x + 16 = 0$

8. $-4x^2 + x - 11 = 0$

For the following quadratic functions, find the x- and y-intercepts. Round your answers to three decimal places.

9. $f(x) = 2x^2 + 5x + 3$

10. $f(x) = 2x^2 - 12x - 14$

11. $f(x) = -2x^2 - 8x - 5$

12. $f(x) = -3x^2 - 5x + 1$

13. $f(x) = x^2 - 4x - 2$

14. $f(x) = x^2 - 5x - 2$

For the following exercises, enter the expressions into your graphing calculator and find the x-intercepts of the function using the graph-intersect method (2nd CALC 5:intersect). Round your answers to the nearest thousandth.

15. $Y_1 = 4x^2 + 3x - 2, Y_2 = 0$

16. $Y_1 = -3x^2 + 8x - 1, Y_2 = 0$

17. $Y_1 = -0.5x^2 + 4.9x - 2.6, Y_2 = 0$

18. $Y_1 = 2.4x^2 + 3.7x - 5.1, Y_2 = 0$

19. To solve the quadratic equation $x^2 + 5x - 7 = 4$, graph these two equations and find the points of intersection. Round solutions to the nearest thousandth.

$$Y_1 = x^2 + 5x - 7$$
$$Y_2 = 4$$

20. To solve the quadratic equation $0.3x^2 + 2x - 4 = 2$, graph the following two equations and find the points of intersection. Round solutions to the nearest thousandth.

$$Y_1 = 0.3x^2 + 2x - 4$$
$$Y_2 = 2$$

21. Abercrombie and Fitch stock had a price P (in dollars) that can be modeled by

$P = 0.2t^2 - 5.6t + 50.2$, where t is the time in months from 1999 to 2001. ($t = 1$ is January 1999). Find the two months in which the price of the stock was $30.

22. A formula for the normal systolic blood pressure for a man age A, measured in mmHg, is given as $P = 0.006A^2 - 0.02A + 120$. Find the age to the nearest year of a man whose normal blood pressure measures 125 mmHg.

23. A flower pot falls off a balcony. It's height (in feet) is given by the formula $f(t) = 100 + 7t - 16t^2$, where t is measured in seconds. How long will it take for the flower pot to hit the ground?

24. A rock is tossed off a cliff and falls to the beach below. It's height (in feet) is given by the formula $f(t) = 122 + 12.5t - 16t^2$, where t is measured in seconds. How long will it take for the rock to hit the beach?

For the following exercises, find the discriminant of the quadratic equation and use it to determine the number and the type of solutions. Then, state the number of x-intercepts of the parabola for the corresponding quadratic function.

25. $0 = -6x^2 - 5x + 2$

26. $0 = 16x^2 - 24x + 9$

27. $0 = 3x^2 - 2x + 8$

28. $0 = 2x^2 + 7x - 4$

29. $0 = x^2 - 10x + 25$

30. $0 = -5x^2 + 3x - 6$

31. $0 = -2x^2 + 5x - 3$

32. $0 = 6x^2 - 2x + 1$

For the following exercises, find all solutions, real or imaginary, using the method of your choice. When needed, round solutions to three decimal places.

33. $2(x-5)^2 - 10 = -82$

34. $0 = 2x^2 - 10x + 4$

35. $x^2 + 7x - 18 = 0$

36. $4x^2 + x = 1$

37. $-3x^2 + 9x = -2$

38. $\frac{1}{2}x^2 + 3x + 1 = 0$

39. $4x^2 - 3x - 1 = 0$

40. $(x-3)^2 + 2 = 0$

41. $-2(x+3)^2 - 6 = 0$

42. $-4 = x^2 + 6x$

43. $(2x-1)(4x+3) + 7 = -5(x^2 - 8)$

44. $(3x-2)(x+5) + 24 = -x^2 - 6$

45. $5x^2 + 48 = 13$

46. $7x^2 + 100 = 37$

47. $0 = x^2 + x - 30$

48. $x^2 + 6x - 55 = 0$

49. $(x+3)(x-2) = 4(x+1)$

50. $3(x+4)^2 = 10$

4.5 Modeling with Quadratic Functions

OVERVIEW

Outside of the college math classroom, you will rarely be asked to find the vertex or the x-intercepts of a quadratic function. In the real world, however, you might very well be asked to find the conditions that produce maximum revenue or minimum cost for a business. You might also want to determine when a projectile will reach its maximum height or when an engine is at the highest point on its power curve. In this section, we will focus on the application of quadratic models to real-world situations. We'll use several features of a graphing calculator to help us solve problems.

In this section, you will learn to:

- Find maximum and minimum values using a graphing calculator.

- Find and interpret both input and output values of a quadratic function in the context of a real-world problem.

- Find a quadratic regression equation using a graphing calculator.

A. USING A GRAPHING CALCULATOR TO FIND MAXIMUM OR MINIMUM VALUES

In Section 4.2, for a quadratic function in standard form, $y = f(x) = ax^2 + bx + c$, we used an algebraic method to determine the coordinates of the vertex of a parabola. Recall that $x = \frac{-b}{2a}$ determines the x-coordinate of the vertex. The y-coordinate is found when we substitute $x = \frac{-b}{2a}$ in the equation and evaluate $f\left(\frac{-b}{2a}\right)$.

For quadratic functions, the vertex represents the point on the curve where either the maximum or minimum value of the function occurs. We can also use the maximum and minimum features on a graphing calculator to find the coordinates of a vertex. Example 1 illustrates this process.

In a real-world problem, the coordinates of the vertex often represent very useful information. The independent variable tells us *when* or *where* the maximum or minimum value of the function occurs. The dependent variable tells us the maximum or minimum value of the function.

▶ *Example 1*

Finding a Maximum Value Using a Graphing Calculator

A company manufactures solar panels. The business manager projects that the profit (in dollars) from making x solar panels per week can be modeled by this formula:

$$P = f(x) = -2.5x^2 + 198x + 10,100$$

How many solar panels should be produced each week in order to maximize profit? What is the maximum profit?

Solution

The leading coefficient, $a = -2.5$, is negative, so the parabola opens downward. The maximum value of the function will be at the vertex. We use a graphing calculator to find the coordinates of the vertex.

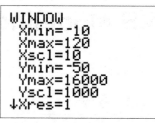

Figure 1. Viewing window

First, enter the equation $y = -2.5x^2 + 198x + 10100$ into the calculator and select a good viewing window. Try [-10, 120, 10] by [-50, 16000, 1000]. See Figure 1.

To approximate the coordinates of the vertex, enter the keystrokes [2nd], [CALC], and [4]. See Figure 2a.

The "maximum" under the [CALC] menu first asks for the left bound. Use your x-axis scaling to help you determine an x-value well to the left of the vertex, say 30, and [ENTER]. It then asks for the right bound. Choose an x-value well to the right of the vertex, say 50, and enter. Finally, the calculator asks for a guess. For the "guess", choose

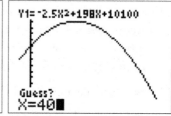

Figures 2a and 2b. Steps for calculating the coordinates of a vertex.

any number between your left and right bounds, say 40, and [ENTER]. See Figure 2b.

In Figure 3, you see that the calculator provides the coordinates of the vertex, (39.6, 14020.4). The value of the independent variable, $x = 39.6$, represents 39.6 solar panels. For practical reasons, we round this to 40. The value of the dependent variable, $y = 14,020.4$ represents the profit of the company in dollars.

Manufacturing 40 solar panels per week will maximize profit. The maximum profit per week is about $14,020.

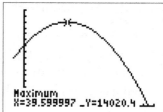

Figure 3. The coordinates of the vertex.

Practice Set A

Practice finding maximum and minimum values using the graphing calculator feature. Round values to three decimal places. Then turn the page to check your solutions.

1. Use a graphing calculator to find the maximum value of $y = -0.4x^2 - 5x - 7$.

2. Use a graphing calculator to find the minimum value of $y = \frac{1}{3}x^2 - \frac{6}{5}x + \frac{1}{2}$.

(Note: In the Calc Menu on your calculator, choose "3: minimum" instead of "4: maximum.")

B. USING AND INTERPRETING A QUADRATIC MODEL

In the next two examples, we apply what we've learned about quadratic functions to make estimates and predictions with realistic quadratic models. We use the graph of a function to help us visualize the relationship between the variables in the model. It's important to be able to interpret solutions correctly within the context of the situation that is modeled.

▶ *Example 2*

Modeling Projectile Motion

A man stands at the edge of a high cliff near the ocean. He throws a rock upward so that it first goes higher than the cliff and then lands on the beach below. The height of the rock, in feet above the beach, is given by $h(t) = -16t^2 + 56t + 124$, where t is time in seconds after the rock is thrown.

   a. Graph the function in a good viewing window.

   b. Find $f(0)$ and $f(3)$ and interpret these values.

   c. When does the rock reach its maximum height? What is the maximum height?

   d. When does the rock hit the beach?

   e. Find t, when $f(t) = 140$. Interpret the solutions.

Solution

a. Enter the equation in the calculator. To choose a good viewing window, we first think about possible values for the input variable t, which represents the time in seconds that the rock is in motion. From personal experience, we estimate that the rock will fall to the beach in a matter of seconds, so we set the x-window to $[0, 8, 1]$.

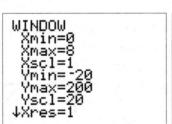 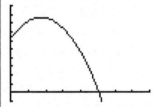

Figure 4. Viewing window and graph.

    The output values represent the height of the rock, so this will include 0 and higher. The value of the constant in the function is 124, so we go above that to ensure we can see the whole graph. We also go below 0 because we want to be able to see the key points on the graph when we find values. Set the y-window to $[-20, 200, 20]$ and graph the function. See Figure 4.

b. Because we already have f graphed, we find $f(0)$ and $f(3)$ using the TRACE mode on our calculator. Press TRACE, type 0, and ENTER. Repeat this process for the input value 2. We find that $f(0) = 124$ and $f(3) = 148$. See Figure 5.

    We interpret these values to mean the rock was first thrown from a height of 124 feet. After 3 seconds, the rock is 148 feet high. Additionally, we can see from the graph that at this time, the rock has

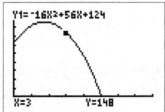

Figure 5. Evaluating $f(3)$ on the graphing screen.

already reached its maximum height and is falling down to the beach. Hopefully nobody is walking on the beach at this moment.

c. The rock reaches the maximum height at the vertex of the parabola. Instead of using the maximum feature on our calculator we choose to use the vertex formula. The input value, t, of the vertex is:

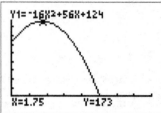

$$t = -\frac{b}{2a} = -\frac{56}{2(-16)} = 1.75$$

The rock reaches a maximum height after 1.75 seconds.

We can easily find the maximum height, which is the y-coordinate of the vertex, using the TRACE mode on the calculator, as you see in Figure 6.

Figure 6. The maximum height of the rock.

The rock reaches a maximum height of 173 feet.

d. To find when the rock hits the beach, we need to determine when the height is zero. Substituting $h(t) = 0$ in the original equation, we have $0 = -16t^2 + 56t + 124$.

We've learned that there is more than one way to solve this equation. To be efficient, this time we use the quadratic formula program on the calculator to solve: $a = -16$, $b = 56$, and $c = 124$. The solutions are $t \approx -1.538$ and $t \approx 5.038$ seconds.

The first solution is negative, which doesn't make sense in the context of the problem. The second solution tells us it takes about 5.038 seconds for the rock to land on the beach. You can verify this on the graphing screen as well.

e. To Find t, when $f(t) = 140$, we substitute in the original equation: $140 = -16t^2 + 56t + 124$.

We could set this equation equal to zero and solve using the quadratic formula, but since we already have the function graphed, let's use the graph-intersect method. Enter $Y_2 = 140$ along with the original function and find the intersection points of the two graphs. Figure 7 shows that the input values are $t \approx 0.314$ and $t \approx 3.186$.

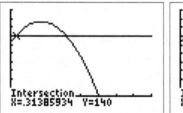 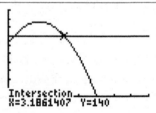

Figure 7. Solutions to the equation $140 = -16x^2 + 56x + 124$

The rock will be 140 feet above the beach at about 0.314 seconds on its way up and again at about 3.186 seconds on its way down.

The next example requires the same set of skills in a slightly different context.

Practice Set A — Answers

1. The vertex is at (–6.25, 8.625). The maximum value is 8.625.

2. The vertex is at (2.80, –3.46). The minimum value is –3.46.

► *Example 3*

Analyzing a Quadratic Model for Profits

A small family farm supplements its income from grain and vegetable crops by farming and selling tilapia fish. Each month, there are overhead costs, and if the fish harvest increases, these costs also rise. This is accounted for in the following function that models the monthly profit in dollars as a function of the number of pounds of fish x the farm sells: $P = f(x) = -0.0001x^2 + 0.6x - 500$.

   a. Find $f(0)$ and interpret this value.

   b. Find the x-intercepts of the graph of this function, and interpret these values.

   c. Find the vertex of the graph of the function and interpret the meaning of both coordinates.

Solution

a. Evaluate the function at $x = 0$:

$$f(0) = -0.0001(0)^2 + 0.6(0) - 500 = -500$$

If the farm sells 0 lbs of fish, it will lose $500. Because of overhead costs, the farm will lose money if it does not sell any tilapia fish.

b. After entering the function into Y_1, it may take some trial and error to come up with a reasonable window setting. [0, 6000, 1000] by [−500, 500, 100] works well. Enter $Y_2 = 0$ and use the [CALC] 5:intersect feature to find the x-intercepts of the graph. Figure 8 shows that the x-intercepts are (1000, 0) and (5000, 0).

If the farm sells 1,000 lbs or 5,000 lbs of fish, it will break even. If it sells between 1,000 and 5,000 lbs, it will make a profit.

c. Use the [CALC] 4:maximum feature on the calculator to find the coordinates of the vertex. The vertex is (3000, 400). See Figure 9.

If the farm sells 3,000 lbs of fish per month they will make a maximum profit of $400.

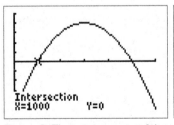 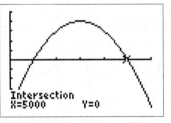 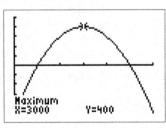

Figure 8. The *x*-intercepts of the graph.   Figure 9.

Practice Set B

3.   A stone is thrown upward from the top of a 112-foot high cliff overlooking the ocean. The stone's height in feet above ocean t seconds after it is thrown can be modeled by the equation $h(t) = -16t^2 + 96t + 112$. Use the model to answer the following questions.

a. Find and interpret $h(2)$.

b. When does the stone reach the maximum height? What is the maximum height?

c. When does the stone land in the ocean?

When you're finished, turn the page to check your answers.

C. FINDING A MODEL USING DATA IN A TABLE AND QUADRATIC REGRESSION

In Section 1.4, we found linear regression equations on the calculator to model data that had the characteristics of a linear function. In Section 2.5, we used exponential regression equations to model data that appeared to grow or decay exponentially. Now we look at data that can be modeled with a quadratic regression equation.

▶ *Example 4*

Fitting a Quadratic Model to Data

A study was done to compare the speed x (in miles per hour) with the mileage y (in miles per gallon) of an automobile. Figure 10 provides the results of the study.

Using the information from Figure 10:

a.   Use a graphing calculator to create a scatter plot of the data.

b.   Use the regression feature of the calculator to find a model that fits the data.

c.   Approximate the speed at which the car gets the greatest mileage.

d.   Predict the mileage when the speed is 85 mph.

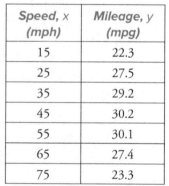

Speed, x (mph)	Mileage, y (mpg)
15	22.3
25	27.5
35	29.2
45	30.2
55	30.1
65	27.4
75	23.3

Figure 10. Speed vs. Mileage.

Solution

a. Use STAT and EDIT to enter the data into Lists 1 and 2 on the calculator. Turn on Plot 1 and graph the data in a good viewing window. The Figure 11 scatter plot shows that the data has a parabolic trend.

b. Use the regression feature — STAT and [CALC] 5:QuadReg — to find the quadratic model. See Figure 12. By rounding the coefficients, we have $y = -0.0084x^2 + 0.771x + 12.87$.

c. Graph the data and the model in the same viewing window as shown in Figure 13. Use the [CALC] 4:maximum feature of the calculator to find the coordinates of the vertex. The vertex is approximately (46, 31). The speed at which the mileage is greatest, then, is about 46 mph. At that speed, the car's mileage is about 31 miles per gallon.

d. Using the trace feature — and making sure that we're tracing the regression model, not the scatter plot — enter "$x = 85$." (You may have to adjust the x-max in the viewing window to 85 or greater.) At 85 miles per hour, the mileage is about 17.5 miles per gallon.

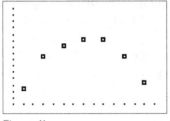
Figure 11.

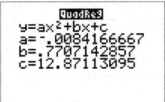
Figure 12.

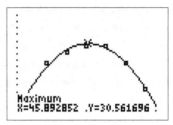
Figure 13.

▶ *Example 5*

Fitting a Quadratic Model to Data

A basketball is dropped from a height of 10 meters. The height of the basketball is recorded at intervals of about 0.1 second. Figure 14 shows the results.

   Use a graphing calculator to find a regression model that fits the data. Then use the model to predict the time when the basketball will hit the ground.

Solution

Begin by entering the data into Lists 1 and 2 on the calculator. Choose a good viewing window and display the scatter plot, as shown in Figure 15.

   From the scatter plot, we see the data has a parabolic trend, so using the regression feature of the calculator, we find the quadratic model, as shown in Figure 16.

   Rounding values of the coefficients, the quadratic model is given by $y = -4.754x^2 - 0.085x + 10.005$ Notice that the y-intercept of the model is approximately (0, 10) which represents the initial height of the basketball.

   To predict when the basketball will hit the ground, we substitute 0 for y in the model and solve the resulting equation for x:

$0 = -4.754x^2 - 0.085x + 10.005$

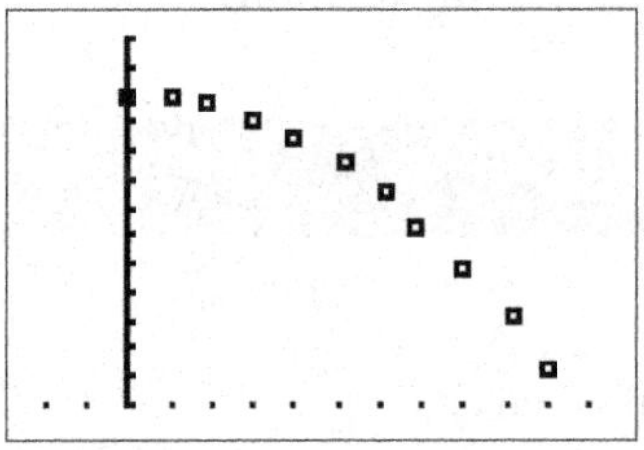

Time, x (seconds)	Height, y (meters)
0	10
0.11	9.952
0.19	9.801
0.30	9.560
0.39	9.215
0.52	8.775
0.61	8.235
0.68	7.594
0.80	6.864
0.92	6.033
1.0	5.107

Figure 14. Height vs. time.

Figure 15.

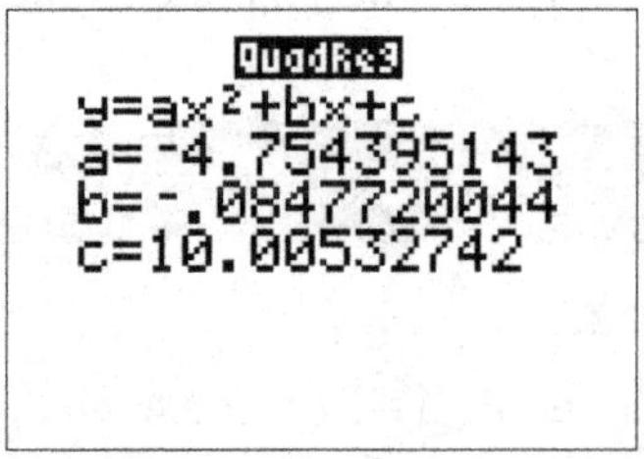

Figure 16.

We use a quadratic formula program on the calculator to solve: $a = -4.754$, $b = -0.085$, and $c = 10.005$. The solutions are $x = -1.460$ and $x = 1.442$. We disregard the negative solution as it is outside the practical domain of the model.

The basketball will hit the ground after about 1.442 seconds.

When fitting an equation model to a set of data, we need to pay close attention to the trend of the data. Sometimes it's obvious from a scatter plot that the trend is linear or quadratic or exponential, but other times it may not be easy to distinguish between models. We can find several models for the data and graph them with the data. Then we visually determine the model that best fits the data. Remember, the best model doesn't have to contain any of the actual data points, but it should come *close* to the data points and be the best overall representation of the trend seen in the data.

▶ *Example 6*

Choosing a Model

The percentage of American college students who are minorities has been increasing since 1970. The table in Figure 17 shows percentages for various years.

Year	Percent
1976	16.2
1980	16.9
1990	20.7
2000	29.6
2005	32.3
2008	34.7

Figure 17. Percentage of American College Students Who Are Minorities (National Center for Education Statistics).

a. Let $y = f(x)$ represent the percentage of American college students who are minorities x years since 1970. Find a linear equation, a quadratic equation, and an exponential equation to model the data. Compare how well the models fit the data.

b. Compare the y-intercepts of the three models and interpret these values.

c. For years before 1970, which of the three models most likely gives the best estimates of the percentages of minority students?

d. Use each model to predict the percentage of American college students who are minorities in 2020.

Solution

a. Enter the data into Lists 1 and 2 in the graphing calculator and graph the scatter plot of the data in a good viewing window. Remember, $x = 0$ corresponds to 1970.

Next, find linear, quadratic, and exponential regression equations and graph each with the scatter plot. Figure 18 presents the regression equations.

Practice Set B Answers

3.

a. $h(2) = 240$. At 2 seconds, the stone is 240 feet above the ocean.

b. The stone reaches its maximum height at 3 seconds. The maximum height is 256 feet.

c. In 7 seconds, the stone will land in the ocean.

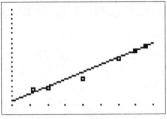
Figure 18a. Linear Model.

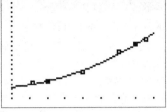
Figure 18b. Quadratic Model.

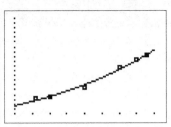
Figure 18c. Exponential Model.

With the coefficients rounded, the regressions are:

Linear: $y = f(x) = 0.6024x + 11.1$
Quadratic: $y = g(x) = 0.0108x^2 + 0.1281x + 14.7$
Exponential: $y = h(x) = 13.4(1.0255)^x$

It appears that the quadratic and exponential models are each a slightly better fit than the linear model.

b. When we round the y-intercept of each model, we find:

Linear: (0, 11.1) In 1970, the model estimates 11.1% minority students.
Quadratic: (0, 14.7) In 1970, the model estimates 14.7% minority students.
Exponential: (0, 13.4) In 1970, the model estimates 13.4% minority students.

Each of these estimates seems reasonable within the context of the problem.

c. To look at the models for the years before 1970, we adjust the viewing window to include negative x and y values. Try: [-40, 40, 10] by [-10, 50, 10]. In Figure 19, we graph all three models on the same screen.

The linear model predicts a negative percentage of minority students before 1950, which does not make sense.

The quadratic model predicts that the percentage of minority students was higher before 1970 and that this percentage decreased before it increased. This seems unlikely given common knowledge about minority access to education in the history of the United States.

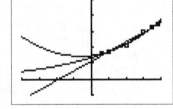
Figure 19. Comparing years before 1970.

The exponential model gives us the most reasonable estimation. Due to the horizontal asymptote at $y = 0$, this model predicts that there were always some — if very few — minorities attending American colleges and that this percentage has always been increasing.

d. To predict the percentage of American college students who are minorities in 2020, we let $x = 2020 - 1970 = 50$ and evaluate each model at this input value:

Linear: $f(50) = 0.6024(50) + 11.1 \approx 41.2$
Quadratic: $g(50) = 0.0108(50)^2 + 0.1281(50) + 14.7 \approx 48.1$
Exponential: $h(50) = 13.4(1.0255)^{50} \approx 47.2$

The linear, quadratic, and exponential models predict 41.2%, 48.1%, and 47.2% minority students respectively. They all seem like reasonable predictions with the linear model showing the smallest growth over time.

Exercises

Use a graphing calculator to solve the following problems. Approximate values to three decimal places where needed.

1. Suppose that the price per unit in dollars of a cell phone production is modeled by $p = 45 - 0.0125x$, where x is in thousands of phones produced, and the revenue represented by thousands of dollars is $R = x \cdot p = x(45 - 0.0125x)$. Write this equation in standard quadratic form and find the production level that will maximize revenue. What is the maximum revenue?

2. Dr. Evil launches a rocket from a submarine. The rocket's height, in meters above sea level, as a function of time in seconds, is given by $h(t) = -4.9t^2 + 229t + 234$. Find the time it takes for the rocket to reach its maximum height. What is the maximum height?

3. A ball is thrown from the top of a building. The ball's height, in meters above ground, as a function of time in seconds, is given by $h(t) = -4.9t^2 + 24t + 8$. How long does it take to reach maximum height and what is that height?

4. Safeco Field in Seattle holds 62,000 Mariners fans. With a ticket price of $11, the average attendance has been 26,000. When the price dropped to $9, the average attendance rose to 31,000. This pattern is accounted for in the following equation which models revenue, R, as a function of the drop in ticket price, x, both in dollars. $R = (11 - x)(26000 + 2500x)$ Write the equation model in standard quadratic form and use it to predict the ticket price that would maximize revenue.

5. A power boat engine has a power curve approximated by $y = -\frac{x^2}{5000} + \frac{21x}{25} - 32$, where x is the number of revolutions per minute (rpm) and y is the horsepower generated. At what number of revolutions per minute is the engine putting out maximum horsepower. What is the maximum horsepower?

6. An engine in a stunt plane has a power curve approximated by $y = -\frac{x^2}{15,000} + \frac{2x}{5} - 12$, where x is the number of revolutions per minute (rpm) and y is the horsepower generated. At what number of revolutions per minute is the engine putting out maximum horsepower. What is the maximum horsepower?

7. The height of a softball is given by $f(t) = -16t^2 + 45t + 6.5$, where y represents the height of the ball in feet and x is the number of seconds that have elapsed since the softball was thrown.

    a. Find $f(0)$ and $f(2)$ and interpret these values.

    b. Find the time it takes for the softball to reach the maximum height. What is the maximum height?

    c. When will the softball hit the ground?

    d. When will the softball be 25 feet in the air?

8. The height of a golf ball is given by $f(t) = -16t^2 + 85t + 0.2$, where y represents the height of the ball in feet and t is the number of seconds that have elapsed since the golf ball was teed off.

    a. Find $f(0)$ and $f(3)$ and interpret these values.

    b. Find the time it takes for the golf ball to reach the maximum height. What is the maximum height?

    c. When will the golf ball hit the ground?

    d. When will the golf ball be 50 feet in the air?

9. The following table shows data collected relating the maximum load in kilograms that a beam can support to the depth of the beam in centimeters.

Depth, x (cm)	Load, y (kg)
10	4,900
12	7,200
14	9,900
16	12,500
18	16,000
20	20,000
22	24,000
24	28,500

   a. Use a graphing calculator to find a quadratic regression model for this data.

   b. Use the model to estimate the load that the beam can support if its depth is 15 cm.

10. The table shows the number y of U.S. Supreme Court cases waiting to be tried for the years 1995 to 2000. (Source: Office of the Clerk, Supreme Court of the U.S.) Let x represent years since 1990, so that $x = 5$ corresponds to 1995.

Year	Cases, y
1995	7565
1996	7602
1997	7692
1998	8083
1999	8445
2000	8965

   a. Use a graphing calculator to find a quadratic regression model for this data.

   b. Use the model to predict the year in which there will be 15,000 U.S. Supreme Court cases waiting to be tried.

11. A new company recorded its sales units versus its profit (or loss) for eight consecutive quarters. See the table below, and note the units sold are in hundreds and the profit is in thousands of dollars.

Units Sold x (hundreds)	Profit, y (thous. $)
0	−30
25	50
40	70
50	70
60	75
75	40
80	30
100	−60

   a. Using a graphing calculator, make a scatter plot

of the data to verify the data has a quadratic trend. Find a quadratic regression model for this data.

   b. Use the model to predict the number of units that should be sold to maximize profit.

   c. Find the x-intercepts of the graph of the model and interpret these values.

12. A new company recorded its production units per day versus its profit (or loss) for those days. See the table below.

Units Sold	Profit ($)
0	−70
10	70
20	140
30	210
40	230
50	170
60	100
70	50

   a. Using a graphing calculator, make a scatter plot of the data to verify the data has a quadratic trend. Find a quadratic regression model for this data.

   b. Use the model to predict the number of units that should be produced to maximize profit.

   c. Find the x-intercepts of the graph of the model and interpret these values.

13. The table shows the amount y (in billions) of dollars spent on books and maps in the United States for the years 1990 to 2000. Let x represent years since 1990.

Year	Amount, y (billion $)
1990	16.5
1991	16.9
1992	17.7
1993	18.8
1994	20.8
1995	23.1
1996	24.9
1997	26.3
1998	28.2
1999	30.7
2000	33.9

   a. Use a graphing calculator to find *both* a linear and a quadratic regression model for this data.

   b. Choose the model that seems to best fit the data. Explain.

   c. Use each model to estimate the sales of books and maps in 2016.

CHAPTER 5

Further Topics in Algebra

In this chapter we cover several different topics that will be useful to the student who intends to enroll in more math or science classes. First we look at direct and inverse variation relationships. In a sealed container, for example, if the pressure on a gas increases, its volume decreases. This is just one case of an inverse variation relationship important to chemists and physicists.

In this chapter, we'll take an introductory look at the patterns found in lists of numbers called sequences. We will also find that we can write formulas to describe these patterns and then use the formulas to solve problems.

5.1 Variation

OVERVIEW

Wally's Used Cars has just offered Shayla a position in sales. The position offers 16% commission on her car sales. For example, if she sells a vehicle for $4,600, she will earn 16% of $4,600, which is $736. Shayla's earnings depend solely on the amount of her sales. As her sales increase, so will her earnings. We say her earnings vary *directly* with her sales.

Each morning before work, Jacob runs 4 miles as part of a healthy lifestyle. The speed at which he runs determines the time it takes to complete the run. If Jacob increases his speed, the time it takes to run 4 miles will decrease. We say the time it takes to complete the run varies *inversely* with the speed.

In this section, we'll look at variation relationships such as the ones described above. You will learn how to:

- ◆ Use a single point to find a direct variation equation or an inverse variation equation.
- ◆ Solve direct variation problems.
- ◆ Solve inverse variation problems.
- ◆ Distinguish between direct and inverse variation models.

A. DIRECT VARIATION

In the example above, Shayla's earnings per vehicle sale can be found by multiplying the sales price of the vehicle, s, by 16%, her commission. The formula $E = 0.16s$ tells us her earnings, E, come from the product of her commission and the sales price of the vehicle.

In Figure 1, we create a table of some possible sales and observe that as the sales price increases, the earnings increase as well. Furthermore, the earnings increase in a predictable way. If we double the sales price of the vehicle from $4,600 to $9,200, we double the earnings from $736 to $1,472.

s, sales price	E = 0.16s	Interpretation
$4,600	$E = 0.16(4,600) = 736$	A sale of a $4,600 vehicle results in $736 earnings.
$9,200	$E = 0.16(9,200) = 1,472$	A sale of a $9,200 vehicle results in $1472 earnings.
$18,400	$E = 0.16(18,400) = 2,944$	A sale of a $18,400 vehicle results in $2944 earnings.

Figure 1.

Figure 2 offers a graphical representation of the equation $E = 0.16s$. As one variable increases, the second variable increases as a multiple of the first. A relationship in which one quantity is a constant multiplied by another quantity is called **direct variation**.

Direct Variation

For $k \neq 0$, if x and y are related by an equation of the form $y = kx$. then we say *y* **varies directly** with x, or y is proportional to x. We call k the **variation constant** or the constant of proportionality. The equation $y = kx$ is called a **direct variation equation**.

The formula we used to represent Shayla's earnings, $E = 0.16s$, is a direct variation equation of the form $y = kx$. We say that earnings vary directly with the sales price of the car. The value $k = 0.16$ is the variation constant in this situation. Here are a few more examples:

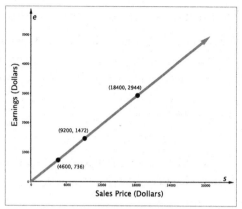

Figure 2. The direct variation equation $E = 0.16s$.

1. For $y = 2.5x$, we say y varies directly with x, with variation constant $k = 2.5$.

2. For $p = 10t$, we say p varies directly with t, with variation constant $k = 10$.

3. For $N = \frac{7}{11}h$, we say N varies directly with h, with variation constant $k = \frac{7}{11}$.

If one variable varies directly with another variable, and if we know one point that lies on the graph, we can find the variation constant k. Once we know the value of k, we can write a direct variation equation.

▶ Example 1

Writing a Direct Variation Equation

If y varies directly with x, and y is 28 when x is 7, find an equation relating x and y.

Solution

The first part of the sentence gives us the general relationship between x and y. The statement "y varies directly with x" is equivalent to the equation $y = kx$.

The second part of the sentence gives us a point on the graph that we use to find the variation constant k. Substitute $x = 7$ and $y = 28$ into the equation and solve for k.

$y = kx$	Start with the general direct variation form.	
$28 = k(7)$	Substitute $x = 7$ and $y = 28$.	
$4 = k$	Divide both sides by 7.	

The variation constant is $k = 4$. The specific direct variation equation is $y = 4x$.

Notice that the equation $y = 4x$ from Example 1 is a linear equation. The graph of $y = 4x$ in Figure 3 is a line with slope $m = 4$ and a y-intercept of $(0, 0)$. In general, the direct variation equation $y = kx$ represents a linear function whose graph is a line that passes through the origin $(0, 0)$. The variation constant k is the slope of this line, or the rate of change in y with respect to x.

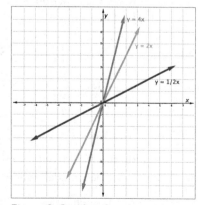

Figure 3. Graphs of $y = kx$.

In Figure 3, we graph three direct variation equations where $k = \frac{1}{2}$, $k = 2$, and $k = 4$. If the value of k is positive, giving us a positive slope, then $y = kx$ is an increasing function. If the value of x increases, the value of y increases.

In general, if we assume y varies directly with x with a positive variation constant k, then if the value of x increases, the value of y increases. If the value of x decreases, the value of y decreases.

Finding and Using a Direct Variation Equation

We're often asked to not only write a direct variation equation but to answer further questions using the equation. First we the given pointon the graph to find the variation constant k and write the equation. Then we use the equation to find the requested value of the variable.

▶ *Example 2*

Using a Direct Variation Equation

The variable p varies directly with q. If $p = 15$ when $q = 35$, find the value of p when $q = 70$. What is the value of q when $p = 54$?

Solution

Represent the direct variation equation as $p = kq$. Next, substitute $p = 15$ and $q = 35$ in the equation and solve for the variation constant k.

$$p = kq \qquad \text{Start with the general direct variation form.}$$
$$15 = k(35) \qquad \text{Substitute } p = 15 \text{ and } q = 35.$$
$$\frac{3}{7} = k \qquad \text{Divide both sides by 35 and reduce the fraction: } \frac{15}{35} = \frac{3}{7}$$

The variation constant is $k = \frac{3}{7} \approx 0.429$. To maintain accuracy, especially for the follow-up questions, we use the fraction, rather than the rounded value, to represent k in our equation:

$$p = \frac{3}{7}q$$

Use the equation to find the value of p when $q = 70$. Substitute $q = 70$ in the equation:

$$p = \frac{3}{7}(70) = 30$$

So $p = 30$ when $q = 70$.

Now use the equation again to find the value of q when $p = 54$. Substitute $p = 54$ in the equation and solve for q:

$$54 = \frac{3}{7}q \qquad \text{Substitute } p = 54 \text{ in the equation.}$$
$$\frac{7}{3} \cdot 54 = \frac{7}{3} \cdot \frac{3}{7}q \qquad \text{Multiply both sides by the reciprocal of } \frac{3}{7} \text{ to solve for } q.$$
$$126 = q$$

So $q = 126$ when $p = 54$.

Variation models are quite common in the sciences, particularly in chemistry and physics. To find and use a direct variation model, follow these steps:

1. Write a general direct variation equation using the variables given.

2. Substitute the given values of the variables in the equation and solve for the constant of variation, k.

3. Use the constant of variation to write the specific direct variation model.

4. Use the equation from Step 3 to make estimates for requested values of the variable.

▶ *Example 3*

Finding and Using a Direct Variation Model

The weight of an object on Mars, M, varies directly with its weight on Earth, E. If an object weighs 100 lbs on Earth, then it only weighs 40 lbs on Mars.

a. *Curiosity* is a car-sized robotic rover exploring Gale Crater on Mars as part of NASA's Mars Science Laboratory mission. This rover weighs almost 2 tons, 4000 lbs., on Earth. How much does it weigh on Mars?

b. *Curiosity* has retrieved and analyzed many rock samples while exploring Mars. If a rock sample weighs 7.2 kg on Mars, how much does it weigh on Earth?

Solution

a. A general direct variation model for this situation will have the form $M = kE$. Given $E = 100$ when $M = 40$, we can solve for k.

$$M = kE \qquad\qquad \text{Start with the direct variation form.}$$
$$40 = k(100) \qquad\qquad \text{Substitute } E = 100 \text{ and } M = 40.$$
$$0.4 = k \qquad\qquad \text{Divide both sides by 100. } \tfrac{40}{100} = 0.4$$

The constant of variation is $k = 0.4$. This time we will use the decimal form of the constant since we do not have to round the value. So the direct variation model is $M = 0.4E$. To find M when $E = 4000$, we substitute in the equation: $M = 0.4(4000) = 3200$. The rover, *Curiosity*, weighs 3200 lbs. on Mars.

b. Substitute $M = 7.2$ in the equation and solve for E.

$$7.2 = 0.4E \qquad\qquad \text{Substitute } M = 7.2.$$
$$18 = E \qquad\qquad \text{Divide both sides by 0.4.}$$

A rock that weighs 7.2 kg on Mars will weigh 18 kg on Earth.

Practice Set A

For each of the following problems, write a direct variation equation, and then find the requested variable values. Turn the page to check your work.

1. If y varies directly with x, and $y = 6.25$ when $x = 3$, find y when $x = 7$.

2. If C varies directly with n, and $C = \tfrac{28}{3}$ when $n = 14$, find n when $C = 18$.

3. The temperature of a certain gas, T, varies directly with its pressure, p. A temperature of 190 K produces a pressure of 50 pounds per square inch (psi). What is the temperature of the gas when the pressure is 65 psi? What pressure will the gas have at 290 K?

B. INVERSE VARIATION

Water temperature in an ocean varies inversely with the water's depth. The formula $T = \frac{14,000}{d}$ gives us the temperature in degrees Fahrenheit at a depth, d, of feet below the ocean's surface at a certain location in the Pacific Ocean. In Figure 4, we have a table of some possible depths and observe that as the depth increases, the water temperature decreases.

Depth, d (ft)	$T = \frac{14,000}{d}$	Interpretation
500	$\frac{14,000}{500} = 28$	At a depth of 500 feet, the water temperature is 28° F.
1000	$\frac{14,000}{1000} = 14$	At a depth of 1,000 feet, the water temperature is 14° F.
2000	$\frac{14,000}{2000} = 7$	At a depth of 2,000 feet, the water temperature is 7° F.

Figure 4.

We notice in the relationship between these variables that as one quantity increases, the other decreases. The two quantities are said to be inversely proportional and each term varies inversely with the other. Inversely proportional relationships are also called **inverse variations**.

We say the water temperature varies inversely with the depth of the water because, as the depth increases, the temperature decreases. Figure 5 offers a graphical representation of this inverse variation.

The formula we used to represent the temperature of the ocean, $T = \frac{14,000}{d}$, is an inverse variation equation with variation constant $k = 14,000$.

> ## Inverse Variation
>
> For $k \neq 0$, if x and y are related by an equation of the form $y = \frac{k}{x}$, then we say y **varies inversely** with x, or y is proportional to x. We call k the **variation constant** or the constant of proportionality. The equation $y = \frac{k}{x}$ is called an **inverse variation equation**.

Here are a few more examples of inverse variation equations:

1. For $y = \frac{5}{x}$, we say y varies inversely with x with variation constant $k = 5$.

2. For $p = \frac{0.22}{w}$, we say p varies inversely with w with variation constant $k = 0.22$.

3. For $P = \frac{7.2}{V}$, we say P varies inversely with V with variation constant $k = 7.2$.

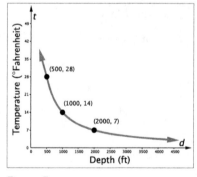

Figure 5.

Finding and Using an Inverse Variation Equation

If one variable varies inversely with another variable and we know one point that lies on the graph, we can find the variation constant k. Once we know the value of k, we can write an inverse variation equation and use it to find requested values.

▶ *Example 4*

Finding and Using an Inverse Variation Equation

If y varies inversely with x, and $y = 9$ when $x = 4$, find y when $x = 7.2$.

Solution

The first part of the sentence gives us the general relationship between x and y. The statement "y varies inversely with x" is equivalent to this equation:

$$y = \frac{k}{x}$$

The second part of the sentence gives us a point on the graph that we use to find the variation constant k. Substitute $x = 4$ and $y = 9$ into the equation and solve for k.

$y = \frac{k}{x}$	Start with the general inverse variation form.
$9 = \frac{k}{4}$	Substitute $x = 4$ and $y = 9$.
$36 = k$	Multiply both sides by 4.

The variation constant is $k = 36$. The specific inverse variation equation is $y = \frac{36}{x}$. To use the equation to find y when $x = 7.2$, substitute $x = 7.2$ in the equation:

$$y = \frac{36}{7.2} = 5$$

So $y = 5$ when $x = 7.2$.

Notice in Example 4 that as the x-value *increased* from $x = 4$, in the first part of the example, to $x = 7.2$, in the second part, the y-value *decreased* from $y = 9$ to $y = 5$.

If we assume y varies inversely with x with a positive variation constant k, then for positive values of x:

- If the value of x increases, the value of y decreases.
- If the value of x decreases, the value of y increases.

▶ *Example 5*

Finding and Using an Inverse Variation Equation

If m varies inversely with b, and $m = 15$ when $b = 7$, find b when $m = 5.25$.

Solution

The general inverse variation equation is $m = \frac{k}{b}$. Substitute $m = 15$ and $b = 7$ into the equation and solve for k.

$15 = \frac{k}{7}$	Substitute $m = 15$ and $b = 7$.
$105 = k$	Multiply both sides by 7.

Practice Set A — Answers

1. $y = 2.25x$, $y = 15.75$
2. $C = \frac{2}{3}n$, $n = 27$
3. $T = 3.8p$, $T = 247$ K, and $p \approx 76.3$ psi

The variation constant is $k = 105$. The specific inverse variation equation is $m = \frac{105}{b}$. To use the equation to find b when $m = 5.25$, substitute $m = 5.25$ in the equation and solve for b.

$$5.25 = \frac{105}{b} \qquad \text{Substitute } m = 5.25.$$

$$5.25b = 105 \qquad \text{Multiply both sides by } b.$$

$$b = 20 \qquad \text{Divide both sides by } 5.25.$$

So $b = 20$ when $m = 5.25$.

We solve for an unknown in an indirect variation problem by following these steps:

1. Identify the input, x, and the output, y.

2. Determine the constant of variation.

3. Use the constant of variation to write an equation for the relationship.

4. Substitute known values into the equation to find the unknown.

Finding and Using an Inverse Variation Model

Inverse variation models, like direct variation models, are common in many fields of science. To find and use an inverse variation model, follow these steps:

1. Write a general inverse variation equation using the variables given.

2. Substitute the given values of the variables in the equation and solve for the constant of variation, k.

3. Use the constant of variation to write the specific inverse variation model.

4. Use the equation from step 3 to make estimates for requested values of the variable.

▶ Example 6

Finding and Using an Inverse Variation Model

The volume, V, of a gas varies inversely with the pressure, P, that the gas exerts on its container. If the volume is 50 cubic feet when the pressure is 48 pounds per square inch (psi), what is the volume if the pressure increases to 80 psi? What pressure would correspond to a volume of 100 cubic feet?

Solution

The general inverse variation equation is $V = \frac{k}{P}$. Substitute $V = 50$ and $P = 48$ in the equation and solve for k.

$$50 = \frac{k}{48} \qquad \text{Substitute } V = 50 \text{ and } P = 48.$$

$$2400 = k \qquad \text{Multiply both sides by } 48.$$

The variation constant is $k = 2400$. The specific inverse variation equation is $V = \frac{2400}{P}$. To find the volume when the pressure is 80 psi, substitute $P = 80$ in the equation:

$$V = \frac{2400}{80} = 30$$

The volume of the gas is 30 cubic feet when the pressure is 80 pounds per square inch.

To find the pressure when the volume is 100 cubic feet, substitute $V = 100$ in the equation and solve for P.

$$100 = \frac{2400}{P} \qquad \text{Substitute } V = 100.$$

$$100P = 2400 \qquad \text{Multiply both sides by } P.$$

$$P = 24 \qquad \text{Divide both sides by } 100.$$

The pressure is 24 pounds per square inch when the volume is 100 cubic feet.

Practice Set B

For each of the following problems, write an inverse variation equation, and then find the requested variable values. Turn the page to check your work.

4. If y varies inversely with x, and $y = 24$ when $x = 3$, find y when $x = 4$.

5. If C varies inversely with F, and $C = 1.6$ when $F = 16$, find F when $C = 2$.

6. The average heart rate, h, in beats per minute (bpm) of most mammals varies inversely with the mammal's average life span, L, in years. A lion has an average heart rate of 63 beats per minute and an average life span of 25 years. What is the average heart rate of an elephant if its life span is about 70 years? Estimate the life span of a mouse whose heart rate is 634 bpm.

C. DIRECT AND INVERSE VARIATION OF A POWER

So far we have described functions in which the dependent variable varies directly or inversely with the independent variable. Some quantities vary directly or inversely with the square of another quantity, or with the cube of another quantity, or with the square root of another quantity, and so on. Here are some examples:

1. $y = kx^4$ y varies directly with the fourth power of x.

2. $r = k\sqrt{A}$ r varies directly with the square root of A.

3. $p = \frac{k}{q^3}$ p varies inversely with the cube of q.

We use "varies directly" to mean the dependent variable is equal to a constant multiplied by a power of the independent variable. We use "varies inversely" to mean the dependent variable is equal to a constant divided by a power of the independent variable.

▶ *Example 7*

Solving a Direct Variation Problem

The quantity y varies directly with the cube of x. If $y = 25$ when $x = 2$, find y when x is 6.

Solution

The general formula for direct variation with a cube is $y = kx^3$. Substitute $x = 2$ and $y = 25$, and then solve for k.

$$y = kx^3 \qquad \text{Start with the general formula.}$$
$$25 = k(2^3) \qquad \text{Substitute } x = 2 \text{ and } y = 25.$$
$$25 = k(8) \qquad \text{Simplify.}$$
$$3.125 = k \qquad \text{Divide both sides by 8.}$$

The variation constant $k = 3.125$. The specific direct variation equation is $y = 3.125x^3$. To find y when $x = 6$, substitute $x = 6$ in the equation and solve for y:

$$y = 3.125(6)^3 = 675$$

So $y = 675$ when $x = 6$.

▶ *Example 8*

Solving an Inverse Variation Problem

The illumination I provided by a car's headlight varies inversely as the square of the distance, d, from the headlight. A headlight produces 60 footcandles (fc) at a distance of 10 ft.

  a. Write an equation that models the relationship between the illumination provided by the headlight and the distance from the headlight.

  b. Use this model to create a table of values displaying the illumination in 5-ft increments, starting at 10 ft.

  c. What will be the illumination at 40 ft?

Solution

a. The general formula for inverse variation with a square is $I = \frac{k}{d^2}$. Substitute $I = 60$ and $d = 10$ in the equation and solve for k.

$$I = \frac{k}{d^2} \qquad \text{Start with the general formula.}$$
$$60 = \frac{k}{10^2} \qquad \text{Substitute } I = 60 \text{ and } d = 10.$$
$$6000 = k \qquad \text{Multiply both sides by } 10^2 = 100.$$

The variation constant is $k = 6000$. So the following equation models this situation:

$$I = \frac{6000}{d^2}$$

b. We'll use a graphing calculator to create a table of values. Enter the equation and set the table to begin at 10 with the change in the independent variable equal to 5. See Figure 6. We observe that as the distance from the headlight increases, the intensity of illumination decreases.

c. We can use the graph or the table to find that at a distance of 40 ft, the illumination is just 3.75 fc.

Figure 6. Table setup and table for $y = \frac{6000}{x^2}$.

Exercises

For the following exercises, write an equation describing the relationship of the given variables.

1. y varies directly with x and when $x = 4$, $y = 80$.

2. y varies directly with x and when $x = 36$, $y = 12$.

3. p varies directly with the cube of q and when $q = 2$, $p = 24$.

4. w varies directly with the square of n and when $n = 5$, $w = 15$.

5. V varies directly with the fourth power of t and when $t = 2.5$, $V = 234.375$.

6. p varies directly with the square root of c and when $c = 36$, $p = 30$.

7. y varies inversely with x and when $x = 4$, $y = 2$.

8. y varies inversely with x and when $x = 3.7$, $y = 46.25$.

9. h varies inversely with the square of d and when $h = 8$, $d = 0.4$.

10. C varies inversely with the cube of L and when $L = 3$, $C = 2$.

11. r varies inversely with the square root of A and when $A = 25$, $r = 3$.

12. Z varies inversely with the cube root of n and when $n = 64$, $Z = 5$.

For the following exercises, use the given information to find the unknown value.

13. y varies directly with x. When $x = 3$, then $y = 12$. Find y when $x = 8.5$.

14. p varies directly with the square of x. When $x = 10$, then $p = 25$. Find p when $x = 6$.

15. D varies directly with the cube of t. When $t = 2$, then $D = 40$. Find t when $D = 135$.

16. B varies directly with the square root of q. When $q = 16$, then $B = 28$. Find q when $B = 700$.

17. y varies inversely with x. When $x = 3$, then $y = 12$. Find y when $x = 4.5$.

18. M varies inversely with the square of n. When $n = 4$, then $M = 3$. Find M when $n = 2$.

19. W varies inversely with the cube of t. When $t = 5$, then $W = 2$. Find t when $W = 31.25$.

20. y varies inversely with the square root of x. When $x = 64$, then $y = 12$. Find y when $x = 36$.

Practice Set B — Answers

4. $y = \frac{72}{x}$; $y = 18$

5. $C = \frac{25.6}{F}$; $F = 12.8$

6. $h = \frac{1575}{L}$; $h = 23$ bpm; $L = 2.5$ years

For the following exercises, assume the variation constant k is a positive number.

21. The variable Q varies directly with b. Describe what happens to the value of Q as b increases.

22. The variable p varies directly with t. Describe what happens to the value of p as t decreases.

23. The variable T varies inversely with m. For positive values of m, what happens to the value of T as m decreases?

24. The variable g varies inversely with h. For positive values of h, what happens to the value of g as h increases?

25. The electrical resistance of a cable of specified length varies inversely with the square of the diameter of the cable. Will a thicker cable have more or less resistance than a narrow cable?

26. The electrical resistance of a wire of specified diameter varies directly with the length of the wire. Will a longer wire have more or less resistance than a short wire?

For the following exercises, use the given information to answer the questions.

27. The distance s that an object falls varies directly with the square of the time, t, of the fall. If an object falls 16 feet in one second, how long for it to fall 144 feet?

28. The velocity v of a falling object varies directly with the time, t, of the fall. If after 2 seconds, the velocity of the object is 64 feet per second, what is the velocity after 5 seconds?

29. The rate of vibration of a string under constant tension varies inversely with the length of the string. If a string is 24 inches long and vibrates 128 times per second, what is the length of a string that vibrates 64 times per second?

30. The volume of a gas held at constant temperature varies inversely with the pressure of the gas. If the volume of a gas is 1200 cubic centimeters when the pressure is 200 millimeters of mercury, what is the volume when the pressure is 300 millimeters of mercury?

31. The weight of an object above the surface of Earth varies inversely with the square of the distance from the center of Earth. If a body weighs 50 pounds when it is 3960 miles from Earth's center, what would it weigh it were 3970 miles from Earth's center?

32. The intensity of light measured in foot-candles varies inversely with the square of the distance from the light source. Suppose the intensity of a light bulb is 0.08 foot-candles at a distance of 3 meters. Find the intensity level at 8 meters.

33. The current C in a circuit varies inversely with its resistance R measured in ohms. When the current in a circuit is 40 amperes, the resistance is 10 ohms. Find the current if the resistance is 12 ohms.

34. The force F (in pounds) needed on a wrench handle to loosen a certain bolt varies inversely with the length L (in inches) of the handle. A force of 35 lbs is needed when the handle is 6 inches long. If a person needs only 20 lbs of force to loosen the bolt, estimate the length of the wrench handle.

5.2 Arithmetic Sequences

OVERVIEW

In this section, we examine patterns in lists of numbers. There is a *wide* variety of pattern types:

2, 4, 8, 16, 32, 64,128	The numbers in the list double.
7, 13, 19, 25, 31, 37, 43	The numbers in the list increase by 6.
1, 4, 9, 16, 25, 36, 49	The numbers in the list are increasing perfect squares.

We will focus on the patterns in lists of numbers that have a correlation to linear functions. In this section, you'll learn to:

♦ Know the meaning of sequence, term, and term number.

♦ Identify arithmetic sequences.

♦ Find a formula for the general term of an arithmetic sequence.

♦ Find a term or term number of an arithmetic sequence.

♦ Use an arithmetic sequence to make estimates and predictions.

A. INTRODUCTION TO SEQUENCES

A section in a movie theater has 10 rows. The first row has 15 seats and each other row has 4 more seats than the row in front of it. So the numbers of seats in the 10 rows are:

15, 19, 23, 27, 31, 35, 39, 43, 47, 51

We call this and any other list of numbers "a sequence."

A **finite sequence** is a sequence that has a last term, such as 51 in the example of the movie theater section. A sequence that does not have a last term is called an **infinite sequence**. For example, the list of positive even numbers —

2, 4, 6, 8, 10, . . .

— is an infinite sequence. We use the ellipsis, three periods in a row, to mean "and so forth." It indicates that we are omitting some numbers in the sequence. When we don't put a final term after the ellipsis, it means that the pattern of numbers continues infinitely.

We also use specific notation to represent the terms in a sequence. The variable a with a number subscript represents a term in a sequence. The value of the subscript is a positive integer that indicates the position of the term in the sequence.

Consider this finite sequence from the movie theater rows:

15, 19, 23, 27, 31, 35, 39, 43, 47, 51

We write the first term of the sequence as a_1, so $a_1 = 15$.

We write the second term of the sequence as a_2 so $a_2 = 19$.

We write third term of the sequence as a_3, so $a_3 = 23$. And so on.

We use the variable n to represent term numbers when working with sequences. Since we cannot have a fractional term position, the values of n are positive intergers.

> **Sequence**
>
> A **sequence** is a function whose domain is the set of positive integers: $n = 1, 2, 3, 4, \ldots$

Consider the infinite sequence of positive even numbers:

2, 4, 6, 8, 10, …

The numbers in this sequence can be found from a formula that defines a function, namely $f(n) = 2n$. We generate the terms of the sequence by finding $f(1) = 2$, $f(2) = 4$, $f(3) = 6$, $f(4) = 8$, and so on. The inputs are positive integers, and the outputs are the terms of the sequence. This justifies a more formal definition of a sequence.

The notation for sequences uses a subscript to indicate the value of the input variable, while the function notation uses parentheses:

Instead of $f(1)$, we write a_1 for the *first term* of the sequence.

Instead of $f(2)$, we write a_2 for the *second term* of the sequence.

Instead of $f(3)$, we write a_3 for the *third term* of the sequence.

Instead of $f(n)$, we write a_n for the *nth term* of the sequence.

The term a_n is called the **nth term of the sequence** or the **general term** of the sequence. The general term is used to define the other terms of the sequence. We call the formula for a_n an **explicit formula**, and it defines the nth term of a sequence using the position of the term. That is, if we know the formula for the general term a_n, then we can find any other term in the sequence. The following examples illustrate how this works.

▶ *Example 1*

Writing the Terms of a Sequence Defined by an Explicit Formula

Write the first five terms and the 25th term of the sequence defined by the explicit formula $a_n = -3n + 8$.

Solution

Substitute $n = 1$ into the formula. Repeat with values 2 through 5, and 25 for n.

$n = 1$ $a_1 = -3(1) + 8 = 5$

$n = 2$ $a_2 = -3(2) + 8 = 2$

$n = 3$ $a_3 = -3(3) + 8 = -1$

$n = 4$ $a_4 = -3(4) + 8 = -4$

$n = 5$ $a_5 = -3(5) + 8 = -7$

$\ldots$

$\ldots$

$n = 25$ $a_{25} = -3(25) + 8 = -67$

The first five terms are 5, 2, –1, –4, and –7. The 25th term is –67.

In Figure 1, we list the sequence values in a table. Figure 2 is a graph of the Figure 1 table.

n	1	2	3	4	5
a_n	5	2	−1	−4	−7

Figure 1.

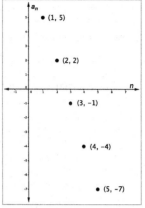

Figure 2.

We can see that this sequence has a linear relationship. However, notice that the points on the graph are not connected by a curve because the domain is the positive integers only.

▶ *Example 2*

Writing the Terms of a Sequence Defined by an Explicit Formula

Write the first five terms and the 50th term of the sequence defined by the explicit formula $a_n = n^2 - 1$.

Solution

Substitute $n = 1$ into the formula. Repeat with values 2 through 5, and 50 for n.

$n = 1$ $a_1 = (1)^2 - 1 = 0$

$n = 2$ $a_2 = (2)^2 - 1 = 3$

$n = 3$ $a_3 = (3)^2 - 1 = 8$

$n = 4$ $a_4 = (4)^2 - 1 = 15$

$n = 5$ $a_5 = (5)^2 - 1 = 24$

...   ...

$n = 50$ $a_{50} = (50)^2 - 1 = 2499$

The first five terms are 0, 3, 8, 15, and 24. The 50th term is 2499.

Practice Set A

Try the following problems. Then turn the page to check your work.

1. Write the first five terms and the 30th term of the sequence defined by the explicit formula $a_n = 5n - 4$.

2. Write the first five terms and the 99th term of the sequence defined by the explicit formula $a_n = \dfrac{n}{n+1}$.

B. DEFINITION OF ARITHMETIC SEQUENCES

Now let's return to the sequence that represents the number of seats in successive rows in a movie theater:

$$15, 19, 23, 27, 31, 35, 39, 43, 47, 51$$

Notice that the difference between any term and the preceding term is 4:

$$19 - 15 = 4$$
$$23 - 19 = 4$$
$$27 - 23 = 4$$
$$\ldots$$
$$51 - 47 = 4$$

> ### Arithmetic Sequence
>
> An **arithmetic sequence** is a sequence where the difference between every pair of consecutive terms is a constant, d. The constant d is called the **common difference**.

With this sequence, we say that 4 is the common difference and that the sequence is arithmetic.

Arithmetic sequences are sequences whose terms change by a constant amount. Each term increases or decreases from the preceding term by the same constant value. This constant value is called the **common difference** of the sequence. Figure 3 shows a sequence with a common difference of -10.

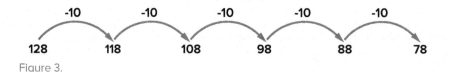

Figure 3.

Figure 4 is another example of an arithmetic sequence. In this case, the common difference is 3. You can choose any term of the sequence and add 3 to find the next term.

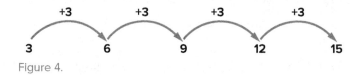

Figure 4.

▶ Example 2

Identifying Arithmetic Sequences

Determine whether the sequence is arithmetic. If it is, find the common difference d.

1. $1, 10, 19, 28, 37, \ldots$
2. $27, 22, 17, 12, 7, \ldots$
3. $2, 6, 18, 54, 162, \ldots$

Solution

Subtract each term from the subsequent term to determine whether a common difference exists.

1. The sequence is arithmetic, because it has a common difference, $d = 9$:

$$10 - 1 = \mathbf{9}$$
$$19 - 10 = \mathbf{9}$$
$$28 - 19 = \mathbf{9}$$
$$37 - 28 = \mathbf{9}$$
$$\cdots$$

2. The sequence is arithmetic, because it has a common difference, $d = -5$.

$$72 - 77 = \mathbf{-5}$$
$$67 - 72 = \mathbf{-5}$$
$$62 - 67 = \mathbf{-5}$$
$$57 - 62 = \mathbf{-5}$$
$$\cdots$$

3. The sequence is *not* arithmetic. We can see from the first two differences that the sequence does not have a common difference: $6 - 2 = 4$ and $18 - 6 = 12$

The graph of each sequence in Example 2 is shown in Figure 5. We can see from the graphs that Figures 5a and 5b are linear, while Figure 5c is *not* linear. Arithmetic sequences have a constant rate of change so their graphs will always be points on a line with slope m equal to the common difference d.

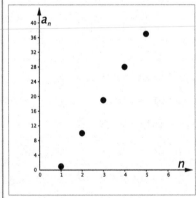

Figure 5a. $d = 9$

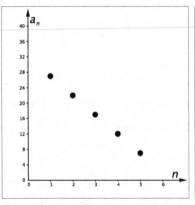

Figure 5b. $d = -5$

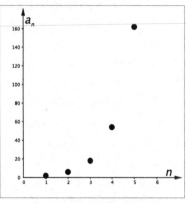

Figure 5c. No common difference

Practice Set B

Determine whether the sequence is arithmetic. If it is, find the common difference d. When you're done, turn the page to check your work.

3. $11, 8, 5, 2, -1, \ldots$ **4.** $1, 3, 6, 10, 15, \ldots$ **5.** $-6, 14, 34, 54, 74, \ldots$

Practice Set A — Answers

1. The first five terms are $1, 6, 11, 16, 21$. The 30th term is 146.

2. The first five terms are $\frac{1}{2}, \frac{2}{3}, \frac{3}{4}, \frac{4}{5}$, and $\frac{5}{6}$. The 99th term is $\frac{99}{100}$.

C. EXPLICIT FORMULA FOR AN ARITHMETIC SEQUENCE

Writing Terms of Arithmetic Sequences

Now that we can recognize an arithmetic sequence, we can also find an explicit formula for the sequence that gives the value of any term when the input is the term number. In general, we can find the terms of an arithmetic sequence by beginning with the first term and adding the common difference repeatedly.

> **Explicit Formula for an Arithmetic Sequence**
>
> An explicit formula for the nth term of an arithmetic sequence with first term a_1 and common difference d is given by
> $$a_n = a_1 + (n-1)d$$

Let's return to the movie theater sequence:

$$15, 19, 23, 27, 31, 35, 39, 43, 47, 51$$

As you've seen, the common difference is 4. We add 4 to the first term, 15, to find the second term, 19. If we add 4 to the first term *two* times, we find the third term, 23. If we add 4 to the first term *three* times, we find the fourth term, 27. And so on.

For an arithmetic sequence with a common difference, d, the terms can also be stated like this:

$$a_1, a_1 + d, a_1 + d + d, a_1 + d + d + d,...$$

We can simplify the general sequence like this:

$$a_1, a_1 + d, a_1 + 2d, a_1 + 3d,...$$

We can use this pattern to find a formula that describes any term, a_n, of an arithmetic sequence:

$a_1 = a_1$

$a_2 = a_1 + d,$ Add d once to the first term to get the second term.

$a_3 = a_1 + 2d,$ Add d twice to the first term to get the third term.

$a_4 = a_1 + 3d$ Add d three times to the first term to get the fourth term.

$a_n = a_1 + (n-1)d$ Add d a total of $(n-1)$ times to the first term to get the nth term.

We see that the nth term of an arithmetic sequence is equal to the first term plus $n - 1$ times the common difference.

▶ *Example 3*

Finding an Explicit Formula for an Arithmetic Sequence

Find an explicit formula for the sequence 15, 19, 23, 27, 31, 35, 39, 43, 47, 51.

Solution

To find a formula for the sequence, we identify the first term, $a_1 = 15$, and the common difference $d = 4$. We substitute these values in the formula $a_n = a_1 + (n-1)d$:

$a_n = 15 + (n-1)(4)$ Substitute $a_1 = 15$ and $d = 4$.

$a_n = 15 + 4n - 4$ Apply the distributive property.

$a_n = 4n + 11$

The explicit formula $a_n = 4n + 11$ describes the sequence. Notice that the form of the equation is linear. The variable is to the first power, and there is a constant rate of change ($m = 4$).

It's always a good idea to check your formula by verifying one or more terms in the sequence:

$a_4 = 4(4) + 11 = 27$, so the fourth term in the sequence is 27.

$a_{10} = 4(10) + 11 = 51$, so the tenth term in the sequence is 51.

The formula checks out.

▶ *Example 4*

Finding a Term of Arithmetic Sequences

Find the 30th term in the sequence $11, 5, -1, -7, -13, \ldots$

Solution

The sequence has a common difference $d = -6$, so the sequence is arithmetic. We use the common difference and the first term, $a_1 = 11$, in the formula $a_n = a_1 + (n - 1)d$ to write the explicit formula for the sequence.

$$a_n = 11 + (n - 1)(-6) \qquad \text{Substitute } a_1 = 11 \text{ and } d = -6.$$
$$a_n = 11 - 6n + 6 \qquad \text{Apply the distributive property.}$$
$$a_n = -6n + 17$$

To find the 30th term, substitute $n = 30$ in the explicit formula: $a_{30} = -6(30) + 17 = -163$. The 30th term of the sequence is -163.

▶ *Example 5*

Finding a Term Number

The number 88 is a term in the sequence $4, 7, 10, 13, \ldots$ What is its term number?

Solution

The sequence has a common difference $d = 3$, so the sequence is arithmetic. We use the common difference and the first term, $a_1 = 4$, in the formula $a_n = a_1 + (n - 1)d$ to write the explicit formula for the sequence.

$$a_n = 4 + (n - 1)(3) \qquad \text{Substitute } a_1 = 4 \text{ and } d = 3.$$
$$a_n = 4 + 3n - 3 \qquad \text{Apply the distributive property.}$$
$$a_n = 3n + 1$$

Practice Set B — Answers

3. The sequence is arithmetic. The common difference $d = -3$.

4. The sequence is not arithmetic because $3 - 1 \neq 6 - 3$.

5. The sequence is arithmetic. The common difference $d = 20$.

To find the term number n for the term 88, substitute $a_n = 88$ in the explicit formula and solve for n.

$$88 = 3n + 1 \qquad \text{Substitute } a_n = 88.$$
$$87 = 3n \qquad \text{Subtract 1 from both sides.}$$
$$29 = n \qquad \text{Divide by 3.}$$

88 is the 29th term in the sequence.

Practice Set C

Now it's your turn. When you're finished, turn the page to check your work.

6. Find an explicit formula for the sequence 27, 20, 13, 6, –1, …

8. Find the term number of the last term in the sequence 2, 11, 20, 29, 38, … , 281

7. Find the 25th term of the sequence –50, –10, 30, 70, 110, …

D. MODELING WITH AN ARITHMETIC SEQUENCE

Some real world situations involve a constant rate of change and an independent variable limited to positive integers. We can model these situations with an arithmetic sequence. The starting value in problem is the first term of the sequence. The constant rate of change is the value of the common difference of the sequence.

As you'll see in Example 6, it can be tricky to align the first term of the sequence with the starting value of the problem. We have to be careful as we define the variable n so that the explicit formula gives the initial value when $n = 1$.

▶ *Example 6*

Modeling with an Arithmetic Sequence

A five-year-old child begins receiving an allowance of $1 each week. Her parents promise her an annual increase of $2.50 per week.

   a. Write a formula for the child's weekly allowance.

   b. What will her allowance be when she is 13 years old?

   c. When will the child receive an allowance of $36 per week?

Solution

We can model the situation with an arithmetic sequence with a first term of 1 and a common difference of 2.5.

a. Let a_n be the amount of the allowance, and let n be the number of years after age 4. Notice that we define n so that $n = 1$ represents the girl at age 5. Now we use the explicit formula for an arithmetic sequence to develop the allowance formula.

$$a_n = 1 + (n-1)(2.5) \qquad \text{Substitute } a_1 = 1 \text{ and } d = 2.5.$$
$$a_n = 1 + 2.5n - 2.5 \qquad \text{Apply the distributive property.}$$
$$a_n = 2.5n - 1.5$$

The explicit allowance formula is $a_n = 2.5n - 1.5$.

b. To find the child's allowance when she is 13 years old, we use $n = 13 - 4 = 9$. This will be the ninth year she receives an allowance from her parents:

$$a_n = 2.5(9) - 1.5 = 21$$

The child's allowance at age 13 will be $21.

c. To determine when the child will receive an allowance of $36 per week, we substitute $a_n = 36$ in the explicit formula and solve for n.

$$36 = 2.5n - 1.5 \qquad \text{Substitute } a_n = 36.$$
$$37.5 = 2.5n \qquad \text{Add 1.5 to both sides.}$$
$$15 = n \qquad \text{Divide both sides by 2.5}$$

Recall that n represents the number of years after age 4, so the child will be 19 years old — and no longer a child! — when she receives an allowance of $36 per week.

▶ Example 7

Modeling with an Arithmetic Sequence

Wooden planks are stacked in layers in a lumber yard so that each layer has 3 less planks than the previous layer. The bottom layer has 52 planks.

    a.   Write a formula for the number of planks in the nth layer of the stack.

    b.   If the top layer of the stack has 10 planks, how many layers are there in the stack?

Solution

a. The situation can be modeled with an arithmetic sequence where the first term is $a_1 = 52$ and the common difference is -3.

$$a_n = 52 + (n-1)(-3) \qquad \text{Substitute } a_1 = 52 \text{ and } d = -3.$$
$$a_n = 52 - 3n + 3 \qquad \text{Apply the distributive property.}$$
$$a_n = -3n + 55$$

The explicit formula is $a_n = -3n + 55$.

b. The top layer has 10 planks, so we substitute a_n = 10 in the explicit formula and solve for n.

$$10 = -3n + 55 \qquad \text{Substitute } a_n = 10.$$
$$-45 = -3n \qquad \text{Subtract 55 from both sides.}$$
$$15 = n \qquad \text{Divide both sides by } -3.$$

The stack of planks has 15 layers.

Exercises

1. What is an arithmetic sequence?

2. How is the common difference of an arithmetic sequence found?

3. How do we determine whether a sequence is arithmetic?

4. Describe how linear functions and arithmetic sequences are similar. How are they different?

In the following exercises, write the first five terms of the sequence. Determine whether or not the sequence is arithmetic. If it is, find the common difference.

5. $a_n = 5 + 12n$

6. $a_n = 8n - 3$

7. $a_n = \dfrac{1}{n+1}$

8. $a_n = 3(2)^n$

9. $a_n = -2 + (n-1)15$

10. $a_n = 5(n+2)$

11. $a_n = 100 - n^2$

12. $a_n = n^3$

In the following exercises, determine whether the sequence is arithmetic. If so, find the common difference d.

13. $0, \frac{1}{2}, 1, \frac{3}{2}, 2, \ldots$

14. $11.4, 9.3, 7.2, 5.1, 3, \ldots$

15. $4, 16, 64, 256, 1024, \ldots$

16. $3, 15, 75, 375, 1875, \ldots$

17. $55, 34, 13, -8, -29, \ldots$

18. $-31, -18, -5, 8, 21, \ldots$

19. $11.4, 9.5, 7.6, 5.7, 3.8, \ldots$

20. $\frac{7}{6}, \frac{6}{5}, \frac{5}{4}, \frac{4}{3}, \frac{3}{2},$

In the following exercises, write an explicit formula for the arithmetic sequence, and then find the 25th term of the sequence.

21. $8, 13, 18, 23, 28, \ldots$

22. $2, 9, 16, 23, 30, \ldots$

23. $27, 20, 13, 6, -1, \ldots$

24. $358, 347, 336, 325, 314, \ldots$

25. $-12, -37, -62, -87, -112, \ldots$

26. $-7, -23, -39, -55, -71, \ldots$

27. $23.5, 25.25, 27, 28.75, 30.5, \ldots$

28. $14.3, 17.1, 19.9, 22.7, 25.5, \ldots$

In the following exercises, find the term number n of the last term of the finite sequence.

29. $7, 12, 17, 22, 27, \ldots, 102$

30. $2, 9, 16, 23, 30, \ldots, 184$

31. $4, 10, 16, 22, 28, \ldots, 1426$

32. $5, 13, 21, 29, 37, \ldots, 533$

33. $-27, -39, -51, -63, -75, \ldots, -999$

34. $-3, -12, -21, -30, -39, \ldots, -390$

35. $675, 650, 625, 600, 575, \ldots -175$

36. $508, 476, 444, 412, 380, \ldots, -772$

37. Is 172 a term in the sequence $2, 9, 16, 23, 30, \ldots$? Explain.

38. Is 1888 a term in the sequence $5, 14, 23, 32, 41, \ldots$? Explain.

With the following exercises, use the information provided to answer the questions.

39. An auditorium has rows of seating that increase in length the further the row is from the stage. The first row has 21 seats, the second row has 24 seats, the third row has 27 seats, the fourth row has 30 seats, and so on.

   a. Let a_n be the number of seats in the nth row. Write an explicit formula to model this situation.

   b. Find a_{12} and interpret this value.

   c. If the last row has 138 seats, how many rows are in the auditorium?

40. A marching band makes a triangle formation during a half-time show at a football game. In this formation there are 3 band members in the first row, 5 members in the second row, 7 members in the third row, 9 members in the fourth row, and so on.

   a. Let a_n be the number of band members in the nth row. Write an explicit formula to model this situation.

   b. Find a_7 and interpret this value.

   c. If there are 49 band members in the last row of the marching formation, how many rows are there?

41. A welder's starting salary is $32,500. Each year she receives a $950 raise.

   a. Let a_n be the welder's salary in the nth year. Write an explicit formula to model this situation.

   b. Find a_5 and interpret this value.

   c. In what year will her salary first exceed $50,000?

42. A teacher's starting salary is $38,000. Each year he receives a $1100 raise.

   a. Let a_n be the teacher's salary in the nth year. Write an explicit formula to model this situation.

   b. Find a_6 and interpret this value.

   c. In what year will his salary be $60,000?

43. A team of 8 people are conducting an experiment at a research station in Antarctica. The team arrived at the station with 225 daily food units. Each team member uses one food unit per day.

   a. Let a_n be the number of food units remaining after the nth day. Write an explicit formula to model this situation.

   b. Find a_{14} and interpret this value.

   c. How many days will the food supply last?

44. Marco has an expense account to use on his world travels. He begins with $19,000. Every Friday, he withdraws $700 to help with his travel expenses.

   a. Let a_n be the balance of the expense account after n weeks. Write an explicit formula to model this situation.

   b. Find a_{15} and interpret this value.

   c. How many weeks will the money last?

Practice Set C — Answers

6. $a_n = -7n + 34$

7. $a_n = 40n - 90$; $a_{25} = 910$

8. $a_n = 9n - 7$; $n = 32$ so 281 is the 32nd term in the sequence.

5.3 Geometric Sequences

OVERVIEW

Many jobs offer an annual cost-of-living increase to keep salaries consistent with inflation. Jamie, for example, is a recent college graduate who is hired as a sales manager and earns an annual salary of $26,000. She's promised a 2% cost of living increase each year. We can find her annual salary in any given year by multiplying her salary from the previous year by 102% (1.02). Her salary will be $26,520 after one year; $27,050.40 after two years; $27,591.41 after three years, and so on. When a salary increases by a constant rate each year, the salary grows by a constant factor.

In this section, we will study sequences that grow in this way. You will learn how to:

+ Identify a geometric sequence.

+ Find a formula for the general term of a geometric sequence.

+ Find a term or term number of a geometric sequence.

+ Use a geometric sequence to make estimates and predictions.

A. DEFINITION OF GEOMETRIC SEQUENCE

In Section 5.2, we worked with arithmetic sequences whose terms increased or decreased by a constant value called the common difference. The common difference was *added* to a term to find the next term. In a geometric sequence, each term increases or decreases by a constant factor called the common ratio. The common ratio is *multiplied* by a term to find the next term. Consider this sequence:

> **Geometric Sequence**
>
> A **geometric sequence** is one in which any term divided by the previous term is a constant r. This constant r is called the **common ratio** of the sequence.

$$1, 6, 36, 216, 1296, \ldots$$

Notice that the ratio of any term to its preceding term is **6**:

$$\frac{6}{1} = 6, \frac{36}{6} = 6, \frac{216}{36} = 6, \frac{1296}{6} = 6, \ldots$$

We call **6** the common ratio, and we call the sequence a geometric sequence. Multiplying any term of the sequence by the common ratio 6 generates the next term, as you see in Figure 1.

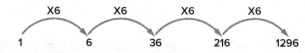

Figure 1. Geometric sequence with common ratio 6.

We can determine if a set of numbers represents a geometric sequence by following these steps:

1. Divide each term by the previous term.
2. Compare the quotients. If they are the same, a common ratio exists and the sequence is geometric.

▶ *Example 1*

Identifying Geometric Sequences

Determine whether the sequence is geometric. If so, find the common ratio r.

1. $7, 21, 63, 189, 567, \ldots$

2. $1, 2, 6, 24, 120, \ldots$

3. $3072, 768, 192, 48, 12, \ldots$

Solution

Divide each term by the previous term to determine whether a common ratio exists.

1. The sequence is geometric, because it has a common ratio of **3**:

$$\frac{21}{7} = 3, \qquad \frac{63}{21} = 3, \qquad \frac{189}{63} = 3, \qquad \frac{567}{189} = 3$$

2. The sequence is not geometric because there is not a common ratio.

$$\frac{2}{1} = 2, \qquad \frac{6}{2} = 3, \qquad \frac{24}{6} = 4$$

3. The sequence is geometric, because it has a common ratio of $\frac{1}{4}$:

$$\frac{768}{3072} = \frac{1}{4}, \qquad \frac{192}{768} = \frac{1}{4}, \qquad \frac{48}{192} = \frac{1}{4}, \qquad \frac{48}{12} = \frac{1}{4}$$

For the third sequence in Example 1, we can write the common ratio in decimal form: $r = \frac{1}{4} = 0.25$. In fraction form, the common ratio reminds us that multiplying a term by $\frac{1}{4}$ gives the same result as dividing a term by 4.

Practice Set A

Determine whether the sequence is geometric. If so, find the common ratio r. When you are finished, turn the page to check your work.

1. $3, 15, 75, 375, 1875, \ldots$ **3.** $3, 9.6, 30.72, 70.656, 162.5088, \ldots$

2. $2.4, 6, 15, 37.5, 93.75, \ldots$ **4.** $720, 360, 180, 90, 45, \ldots$

B. EXPLICIT FORMULA FOR A GEOMETRIC SEQUENCE

Now that we can identify a geometric sequence, we learn how to find any term of a geometric sequence if we know the first term and the common ratio. We can find the terms of a geometric sequence by beginning with the first term and multiplying by the common ratio repeatedly. For example, if the first term of a geometric sequence is $a_1 = 2$ and the common ratio is $r = 4$, we find the second term by multiplying $2 \cdot 4$ to get 8, we find the third term by multiplying $2 \cdot 4 \cdot 4$ to get 32, and so on:

$$
\begin{aligned}
a_1 &= 2 \\
a_2 &= 2 \cdot 4 & &= 8 \\
a_3 &= 2 \cdot 4 \cdot 4 & &= 32 \\
a_4 &= 2 \cdot 4 \cdot 4 \cdot 4 & &= 128
\end{aligned}
$$

The first four terms are 2, 8, 32, 128.

In general, if a_1 is the first term of a geometric sequence and r is the common ratio, the sequence will look like this:

$$a_1, a_1 \cdot r, a_1 \cdot r \cdot r, a_1 \cdot r \cdot r \cdot r, \dots$$

Using exponents, we have this:

$$a_1, a_1 r, a_1 r^2, a_1 r^3, \dots$$

We use the pattern to find a formula that describes the general term a_n of a geometric sequence.

> **Explicit Formula for a Geometric Sequence**
>
> The nth term of a geometric sequence is given by the explicit formula:
>
> $$a_n = a_1 r^{n-1}$$
>
> where a_1 is the first term and r is the common ratio of the sequence.

$$a_1 = a_1$$
$$a_2 = a_1 r \qquad \text{Multiply } a_1 \text{ by } r \text{ once to get the second term.}$$
$$a_3 = a_1 r^2 \qquad \text{Multiply } a_1 \text{ by } r \text{ twice to get the third term.}$$
$$a_4 = a_1 r^3 \qquad \text{Multiply } a_1 \text{ by } r \text{ three times to get the fourth term.}$$
$$a_n = a_1 r^{n-1} \qquad \text{Multiply } a_1 \text{ by } r \text{ a total of } (n-1) \text{ times to get the } n\text{th term.}$$

▶ Example 2

Finding an Explicit Formula

Find a formula that describes the terms of the sequence.

1. 2, 6, 18, 54, 162, …   **2.** 32, 16, 8, 4, 2, …

Solution

1. The sequence has a common ratio of $r = 3$, so the sequence is geometric. Substitute $a_1 = 2$ and $r = 3$ into the explicit formula $a_n = a_1 r^{n-1}$:

$$a_n = 2(3)^{n-1}$$

We can verify the explicit formula by entering $y = 2(3)^{x-1}$ in a graphing calculator and checking with the table feature that the first five terms are 2, 6, 18, 54, and 162. See Figure 2.

2. The sequence has a common ratio of $r = \frac{1}{2}$, so the sequence is geometric. Substitute $a_1 = 32$ and $r = \frac{1}{2}$ into the explicit formula $a_n = a_1 r^{n-1}$:

$$a_n = 32\left(\frac{1}{2}\right)^{n-1}$$

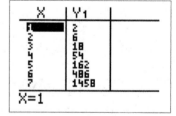

Figure 2. Verify the formula.

Notice that an explicit formula for a geometric sequence is in a form that is similar to the form of the exponential functions that we studied in Chapter 2. For both forms —

$$a_n = a_1 r^{n-1} \text{ and } y = ab^x$$

— the base is a constant number and the variable is in the exponent. One difference is we subtract 1 from the input in the geometric sequence formula to ensure the initial value is the first term of the sequence corresponding to $n = 1$. The exponential function formula has an initial value, a, that corresponds to $x = 0$.

The other main difference between the two forms is that the domain of the formula for geometric sequences is limited to positive integers: $n = 1, 2, 3, 4, \ldots$

The first formula from Example 2, $a_n = 2(3)^{n-1}$, describes an exponential function whose only inputs are positive integers and whose outputs are the terms of the sequence. The points $(1, 2)$, $(2, 6)$, $(3, 18)$, $(4, 54)$, and $(5, 162)$ are on the graph in Figure 3a. Because the base is 3, which is greater than 1, the graph has the shape of an exponential growth curve. We don't connect the dots because we can't have fractional term numbers.

Similarly, the second formula in Example 2, $a_n = 32\left(\frac{1}{2}\right)^{n-1}$, describes an exponential function with positive integer inputs. The points $(1, 32)$, $(2, 16)$, $(3, 8)$, $(4, 4)$, and $(5, 2)$ are on the graph in Figure 3b. Because the base is $\frac{1}{2}$, which is less than 1, the graph has the shape of an exponential decay curve. Again, we do not connect the dots.

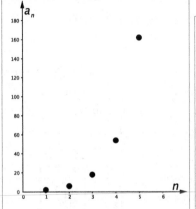

Figure 3a. The first five terms of the sequence $a_n = 2(3)^{n-1}$.

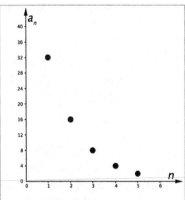

Figure 3b. The first five terms of the sequence $a_n = 32\left(\frac{1}{2}\right)^{n-1}$.

Practice Set B

Write an explicit formula for each geometric sequence. When you're finished, turn the page to check your work.

5. $\frac{1}{8}, \frac{1}{4}, \frac{1}{2}, 1, 2, \ldots$

6. $2000, 400, 80, 16, 3.2, \ldots$

7. $1, 3, 9, 27, 81, \ldots$

8. $90, 9, 0.9, 0.09, 0.009, \ldots$

Practice Set A — Answers

1. geometric, $r = 5$

2. geometric, $r = 2.5$

3. not geometric

4. geometric, $r = \frac{1}{2}$

C. FINDING A TERM OR A TERM NUMBER OF A GEOMETRIC SEQUENCE

For a geometric sequence, we can use the explicit formula $a_n = a_1(r)^{n-1}$ to find a term of the sequence. We input the term number for n in the formula and compute the value of the term. We can also use the formula to solve for a term number, n, when we know a term in the sequence.

▶ *Example 3*

Finding a Term

1. Find the 15th term of the sequence 5, 10, 20, 40, 80, …

2. Find the 12th term of the sequence 15625, 3125, 625, 125, 5, . . .

Solution

1. The sequence has a common ratio, $r = 2$, so the sequence is geometric. The first term is $a_1 = 5$. We substitute these values in the formula for the sequence: $a_n = 5(2)^{n-1}$.

   To find the 15th term of the sequence, we substitute $n = 15$ in the formula:

$$a_{15} = 5(2)^{15-1} = 5(2)^{14} = 81,920$$

To verify our work, we enter $y = 5(2)^{x-1}$ in a graphing calculator and check that the first 5 terms are 5, 10, 20, 40, and 80. Then we check that the 15th term is 81,920. See Figure 4.

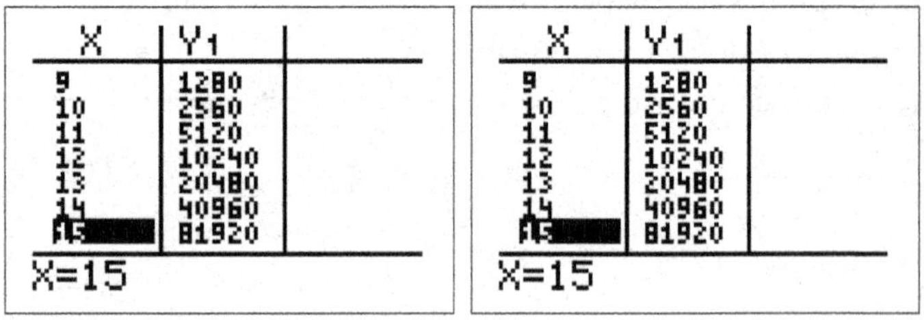

X	Y1
9	1280
10	2560
11	5120
12	10240
13	20480
14	40960
15	81920

X=15

X	Y1
9	1280
10	2560
11	5120
12	10240
13	20480
14	40960
15	81920

X=15

Figure 4. Verify the work.

2. The sequence has a common ratio, $r = \frac{1}{5}$, so it's geometric. The first term is $a_1 = 15,625$. We substitute these values in the formula for the sequence: $a_n = 15,625\left(\frac{1}{5}\right)^{n-1}$.

   To find the 12th term of the sequence, we substitute $n = 12$ in the formula:

$$a_{12} = 15,625\left(\frac{1}{5}\right)^{12-1} = 15,625\left(\frac{1}{5}\right)^{11} = 3.2 \times 10^{-4}$$

The 12th term is expressed in scientific notation. See Figure 5 for a calculator display of this value. In decimal form the 12th term of this sequence is 0.00032.

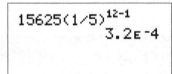

Figure 5. Finding the 15th term of the sequence.

In the next example, we are given a term in a geometric sequence and asked to find its term number. That requires us to solve an exponential equation for the variable n. We use the properties of logarithms we studied in Chapter 3 to help us solve for n.

▶ *Example 4*

Finding a Term Number

The number 885,735 is a term in the sequence 5, 15, 45, 135, 405, ... What is its term number?

Solution

The sequence has a common ratio, $r = 3$, so the sequence is geometric. We note that the first term is $a_1 = 5$ and write the explicit formula for the sequence: $a_n = 5(3)^{n-1}$.

To find the term number, we substitute $a_n = 885,735$ in the formula and solve for n.

$$885735 = 5(3)^{n-1} \qquad \text{Substitute } a_n = 885,735.$$

$$177{,}147 = 3^{n-1} \qquad \text{Divide both sides by 5.}$$

$$\log(177{,}147) = \log(3^{n-1}) \qquad \text{Take the logarithm of both sides.}$$

$$\log(177{,}147) = (n-1)\log(3) \qquad \textbf{Apply the power rule for logarithms.}$$

$$\frac{\log(177{,}147)}{\log(3)} = n - 1 \qquad \text{Divide both sides by } \log(3).$$

$$\frac{\log(177{,}147)}{\log(3)} + 1 = n \qquad \text{Add 1 to both sides.}$$

$$12 = n \qquad \text{Compute.}$$

We find that 885,735 is the 12th term of the sequence. We can use a graphing calculator to verify our work.

Practice Set C

Try the following problems and then turn the page to check your work.

9. Find the 9th term of the sequence 1536, 384, 96, 24, …

10. Find the 11th term of the sequence 5, 30, 180, 1080, 6480, …

11. The number 32 is a term in the sequence 19531.25, 7812.5, 3125, 1250, …. What is its term number?

Practice Set B — Answers

5. $a_n = \frac{1}{8}(2)^{n-1}$

6. $a_n = 2000\left(\frac{1}{5}\right)^{n-1}$ or $a_n = 2000(0.2)^{n-1}$

7. $a_n = (3)^{n-1}$

8. $a_n = 90\left(\frac{1}{10}\right)^{n-1}$ or $a_n = 90(0.1)^{n-1}$

D. MODELING WITH A GEOMETRIC SEQUENCE

Some real world situations involve quantities that increase or decrease by a constant factor and an independent variable limited to positive integers. These situations can be modeled with a geometric sequence. The starting value in the problem is the first term of the sequence. The constant factor is the value of the common ratio of the sequence.

▶ *Example 5*

Modeling with a Geometric Sequence

A pendulum swings 15 feet left to right on its first swing. On each swing following the first, the pendulum swings $\frac{4}{5}$ of the distance of the previous swing. Let a_n represent the distance the pendulum travels on the nth swing.

   a. Find a formula that describes a_n.

   b. Find a_5 and interpret this value.

   c. How many swings does it take before the distance the pendulum travels is less than 1 foot?

Solution

a. The distance of any swing of the pendulum is multiplied by a factor of $\frac{4}{5}$ to find the distance of the next swing. So a_n is a geometric sequence with a common ratio $r = \frac{4}{5}$ and first term $a_1 = 15$. We substitute these values in the formula $a_n = a_1 r^{n-1}$:

$$a_n = 15\left(\tfrac{4}{5}\right)^{n-1}$$

b. To find a_5, we substitute $n = 5$ in the formula:

$$a_5 = 15\left(\tfrac{4}{5}\right)^{5-1} = 15\left(\tfrac{4}{5}\right)^{4} = 6.144$$

  The pendulum travels 6.144 feet on the fifth swing.

c. To find when the distance is than 1 foot, we let $a_n = 1$ in the formula and solve for n.

$$1 = 15\left(\tfrac{4}{5}\right)^{n-1} \qquad \text{Substitute } a_n = 1.$$

$$\tfrac{1}{15} = \left(\tfrac{4}{5}\right)^{n-1} \qquad \text{Divide both sides by 15.}$$

$$\log\left(\tfrac{1}{15}\right) = \log\left(\left(\tfrac{4}{5}\right)^{n-1}\right) \qquad \text{Take the logarithm of both sides.}$$

$$\log\left(\tfrac{1}{15}\right) = (n-1)\log\left(\tfrac{4}{5}\right) \qquad \textbf{Apply the power rule for logarithms.}$$

$$\frac{\log\left(\tfrac{1}{15}\right)}{\log\left(\tfrac{4}{5}\right)} = n - 1 \qquad \text{Divide both sides by } \log\left(\tfrac{4}{5}\right).$$

$$\frac{\log\left(\frac{1}{15}\right)}{\log\left(\frac{4}{5}\right)} + 1 = n \qquad \text{Add 1 to both sides.}$$

$$13.1359 \approx n \qquad \text{Compute.}$$

In this situation, n is the number of swings, so we round our answer to the least integer greater than 13.1359. Then $n = 14$ represents 14 swings of the pendulum.

To verify our work, we use the formula $a_n = 15\left(\frac{4}{5}\right)^{n-1}$ to evaluate a_{13} and a_{14}.

$$a_{13} = 15\left(\frac{4}{5}\right)^{13-1} \approx 1.03 \qquad \text{On the 13th swing, the pendulum travels about 1.03 feet.}$$

$$a_{14} = 15\left(\frac{4}{5}\right)^{14-1} \approx 0.82 \qquad \text{On the 14th swing, the pendulum travels about 0.82 feet.}$$

This verifies that on the 14th swing, the pendulum travels less than 1 foot.

▶ Example 6

Comparing Geometric and Arithmetic Models

Sarah has a new job with a large company, and her employer offers her two choices for structuring her yearly salary:

♦ *Plan A:* A starting salary of $30,000 with a 3.2% raise in her salary each year.

♦ *Plan B:* A starting salary of $34,000 with an $800 raise in her salary each year.

Let a_n be Sarah's salary in dollars in the nth year. For both salary plans, complete the following:

a. Find a formula that describes a_n.

b. Predict her salary, to the nearest whole dollar, in the 5th year and in the 10th year.

c. If she stays with the same company for 32 years, what will her salary be the year she retires?

d. What advice could you give Sarah regarding which plan to choose?

Solution

a. Start by finding a formula that describes a_n for both plans.

Plan A: The sequence is geometric with first term $a_1 = 30,000$ and common ratio $r = 1 + 0.032 = 1.032$. We substitute these values in the formula $a_n = a_1 r^{n-1}$:

$$a_n = 30000(1.032)^{n-1}$$

Plan B: The sequence is arithmetic with first term $a_1 = 34,000$ and common difference $d = 800$. We substitute these values in the formula $a_n = a_1 + (n-1)d$:

$$a_n = 34,000 + (n-1)800, \text{ which is equivalent to } a_n = 800n + 33,200$$

Practice Set C — Answers

9. 0.0234375	**10.** 302,330,880	**11.** $n = 8$

b. Enter the formulas in a graphing calculator. Use a table to find a_5 and a_{10}:

$$Y_1 = 30000(1.032)^{x-1}$$
$$Y_2 = 800x + 33,200$$

Plan A: As Figure 6 shows, in the 5th year, Sarah's salary is $34,028. In the 10th year, it's $39,833.

Plan B: In the 5th year, her salary is $37,200. In the 10th year, it's $41,200.

X	Y1	Y2
5	34028	37200
6	35117	38000
7	36241	38800
8	37401	39600
9	38597	40400
10	39833	41200
11	41107	42000

X=5

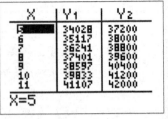

Figure 6.

c. To find Sarah's salary in the 32nd year, use the calculator table to find a_{32} for both plans. See Figure 7.

Plan A: In the 32nd year, Sarah's salary is $79,651.

Plan B: In the 32nd year, her salary is $58,800.

X	Y1	Y2
26	65935	54000
27	68045	54800
28	70222	55600
29	72469	56400
30	74788	57200
31	77181	58000
32	79651	58800

X=32

Figure 7.

d. We notice from part b that Sarah's salary will be higher in the 5th year and the 10th year if she chooses Plan B. From part c we see that her salary in the 32nd year is quite a bit higher with Plan A. As we scroll through the table on the calculator, we can see that Plan B has higher yearly salaries than Plan A until the 12th year. After that, Plan A always has higher salaries. See Figure 8.

We advise Sarah to choose Plan B if she only plans to stay with the company for 12 years or less. We advise her to choose Plan A if she plans to stay with the company for the rest of her working life.

X	Y1	Y2
26	65935	54000
27	68045	54800
28	70222	55600
29	72469	56400
30	74788	57200
31	77181	58000
32	79651	58800

X=32

Figure 8a.

X	Y1	Y2
6	35117	38000
7	36241	38800
8	37401	39600
9	38597	40400
10	39833	41200
11	41107	42000
12	42423	42800

X=12

Figure 8b.

Exercises

1. What is a geometric sequence?

2. How is the common ratio of a geometric sequence found?

3. Describe how exponential functions and geometric sequences are similar. How are they different?

4. What is the difference between an arithmetic sequence and a geometric sequence?

For the following exercises, determine whether the sequence is arithmetic, geometric, or neither. If the sequence is arithmetic, write the common difference. If the sequence is geometric, write the common ratio.

5. 1, 3, 9, 27, 81, …

6. 14, 27, 40, 53, 66, …

7. 200, 180, 160, 140, 120, …

8. 2, 10, 60, 420, 3360, …

9. 6, 8, 11, 15, 20, …

10. 0.8, 4, 20, 100, 500, …

11. $1, \frac{1}{2}, \frac{1}{4}, \frac{1}{8}, \frac{1}{16}, \ldots$

12. 5, 5.2, 5.4, 5.6, 5.8, …

13. $a_n = 9n - 5$

14. $a_n = 75(0.4)^{n-1}$

15. $a_n = n^2 + 4$

16. $a_n = n^3 - 1$

For the following exercises, write an explicit formula for each geometric sequence.

17. 7, 28, 112, 448, 1792, …

18. 0.6, 6, 60, 600, 6000, …

19. $1, \frac{4}{5}, \frac{16}{25}, \frac{64}{125}, \ldots$

20. $2, \frac{1}{3}, \frac{1}{18}, \frac{1}{108}, \ldots$

21. 768, 192, 48, 12, 3, …

22. 1250, 250, 50, 10, 2, …

23. 10, 70, 490, 3430, 24010, …

24. 1, 8, 64, 512, 4096, …

For the following exercises, find the indicated term of the geometric sequence. If needed, write the result in scientific notation $N \times 10^k$, with N rounded to three decimal places.

25. Find the 16th term of the sequence 6, 12, 24, 48, …

26. Find the 13th term of the sequence 2, 6, 18, 54, …

27. Find the 15th term of the sequence 1280, 320, 80, 20, …

28. Find the 14th term of the sequence 5625, 1125, 225, 45, …

29. Find the 25th term of the sequence 3.2, 9.6, 28.8, 86.4, …

30. Find the 24th term of the sequence 1.4, 3.5, 8.75, 21.875, …

For the following exercises, the given number is a term in the geometric sequence that follows. Find the term number of that term. Use a graphing calculator table to verify your result.

31. 25,165,824; 6, 24, 96, 384, …

32. 177147; 1, 3, 9, 27, …

33. 2,470,629; 3, 21, 147, 1029, …

34. 109,375; 0.00224, 0.0112, 0.056, 0.28, …

35. 0.01953125; 640, 320, 80, 5, …

36. 0.46875; 80, 240, 120, 60, …

37. 768; 0.046875, 0.09375, 0.1875, 0.375, …

38. 28,697,814; 2, 6, 18, 54, …

For the following exercises, use the given information to answer the questions.

39. A ball is dropped from a height of 20 feet. Each time it bounces it returns to $\frac{7}{8}$ of the height it fell from. Let a_n represent the maximum height of the ball on the nth bounce.

a. Find a formula that describes a_n.

b. What is the maximum height of the ball on the 9th bounce?

40. A ball is dropped from a height of 8 meters. Each time it bounces it returns to $\frac{4}{5}$ of the height it fell from. Let a_n represent the maximum height of the ball on the nth bounce.

   a. Find a formula that describes a_n.

   b. What is the maximum height of the ball on the 9th bounce?

41. You go to work for a company that pays $0.01 the first day, $0.02 the second day, $0.04 the third day, and so on. If the daily wage keeps doubling, to the nearest dollar what will your income be on the 30th day?

42. You go to work for a company that pays $5 the first week, $15 the second week, $45 the third week, and so on. If the weekly pay keeps tripling, what will your pay be for the 12th week?

43. A sealed glass vessel has a volume of 500 cubic centimeters. With each cycle, a vacuum pump removes one-third of the air in the vessel and two-thirds remains. Let a_n represent the volume of air remaining in the vessel after n cycles of the vacuum pump.

   a. Find a formula that describes a_n.

   b. Find a_4 and interpret this value.

   c. How many cycles of the vacuum pump are needed before there is less than 5 cubic centimeters of air in the vessel?

44. A tire swing attached to a tree branch swings with an arc length of 18 feet on its first swing. On each swing following the first, the tire's arc length is $\frac{3}{4}$ of the length of the previous swing. Let a_n represent the arc length on the nth swing.

   a. Find a formula that describes a_n.

   b. Find a_6 and interpret this value.

   c. How many swings does it take before the arc length is less than 1 foot?

45. Aaron and Beatriz both just graduated from college and received job offers. Aaron was offered a starting salary of $40,000 with a 3.75% raise per year. Beatriz was offered a starting salary of $45,000 with a $1000 raise per year. Let a_n be the salary in dollars in the nth year. For both Aaron and Beatriz complete the following:

   a. Find a formula that describes a_n.

   b. Predict the salary, to the nearest whole dollar, in the 4th year and in the 12th year.

   c. If they each stay with the same company for 30 years, what will their salary be?

   d. In what year does Aaron's salary first become greater than Beatriz's?

SOLUTIONS

Odd-Numbered Exercises

Chapter 1: Graphs and Linear Functions

1.1 QUALITATIVE GRAPHS

1. Figure 7d.

3. Figure 7a.

5. Figure 8b.

7. Figure 8c.

9. p is independent, and t is dependent.

11. h is independent, and T is dependent.

13. d is independent, and h is dependent.

15.

17.

19.

21.

23.

25.

27.

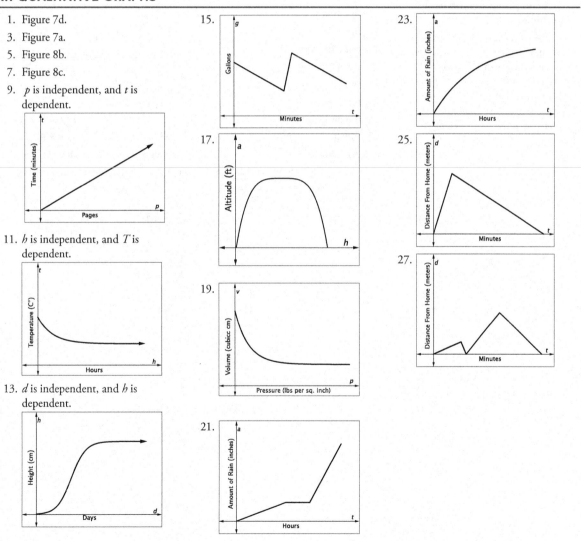

29. The scenarios will vary, but in general, one should describe a dependent variable y that increases as the independent variable x increases. More specifically, the rate of increase slows mid-graph (or mid-scenario) and picks up again after that.

31. The scenarios will vary, but in general, one should describe a dependent variable y that increases then decreases as the independent variable x increases. An example might be the height of a toy rocket, represented by y, over a time interval represented by x. The rocket is launched, goes up, and then falls to the ground.

1.2 FUNCTIONS

1. A relation is a set of ordered pairs. A function is a special kind of relation in which no two ordered pairs have the same first coordinate.

3. When a vertical line intersects the graph of a relation more than once, that indicates that for that input there is more than one output. At any particular input value, there can be only one output if the relation is to be a function.

5. function

7. not a function

9. function

11. function

13. function

15. function

17. not a function

19. not a function

21. function

23. function

25. not a function

27. function

29. not a function

31. Domain: $-2 \le x \le 3$ or $[-2, 3]$; Range: $-4 \le y \le 4$ or $[-4, 4]$

33. Domain: $-5 \le x < 2$ or $[-5, 2)$; Range: $-5 \le y \le 3$ or $[-5, 3]$

35. Domain: $x \ge -2$ or $[-2, \infty)$; Range: $y \ge -3$ or $[-3, \infty)$

37. Domain: *all Real Numbers* or $(-\infty, \infty)$; Range: $y \le 4$ or $(-\infty, 4]$

39. Domain: $1998 \le x \le 2008$ or $[1998, 2008]$; Range: $81 \le y \le 139$ or $[81, 139]$

41. Domain: $[0, 9.8)$; Range: $[0, 225)$

43. Domain: $[-12, \infty)$; Range: $[-\infty, -7)$

45. a) $y = x^2 + 2$

b)

x	−3	−2	−1	0	1	2	3
y	11	6	3	2	3	6	11

c)

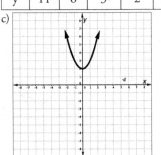

d) Domain: $(-\infty, \infty)$; Range: $[2, \infty)$

47. a) Multiply the input by -2 and add 5 to obtain the output.

b)

x	−3	−2	−1	0	1	2	3
y	11	9	7	5	3	1	−1

c)

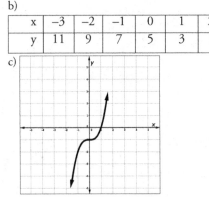

d) Domain: $(-\infty, \infty)$; Range: $(-\infty, \infty)$

1.3 FINDING EQUATIONS OF LINEAR FUNCTIONS

1. a) Terry starts at an elevation of 3000 feet and descends 70 feet per second.

   b)

x	0	10	20	30	40
y	3000	2300	1600	900	200

   c)

   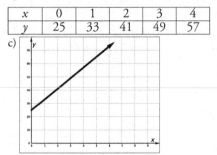

3. a) $y = 8x + 25$

   b)

x	0	1	2	3	4
y	25	33	41	49	57

   c)

5. Increasing

7. Decreasing

9. Decreasing

11. Increasing

13. Decreasing

15. $m = 250.25 \approx 250$ The total number of visitors to the museum increases by approximately 250 visitors per month.

17. $m = -1.5$ The scuba diver *descends* at a rate of 1.5 feet per second.

19. $y = 42.50x + 120$ The slope, $m = 42.50$, indicates the plumber charges a rate of 42.50 dollars per hour. The y-intercept (0, 120) indicates the plumber's initial charge is 120 dollars. If the plumber works for $x = 5.5$ hours, then he will charge $353.75.

21. $y = -13x + 4500$ The slope, $m = -13$, indicates that the volume of grain in the silo decreases at a rate of 13 cubic meters per minute. The y-intercept (0, 4500) indicates the initial amount of grain is 4500 cubic meters. After 4 hours, $x = 240$ minutes and there will be 1380 cubic meters of grain remaining in the silo.

23. $y = -3.6x + 100$ The slope, $m = -3.6$, indicates the equipment descends at a rate of 3.6 meters per minute. The y-intercept (0, 100) indicates the initial distance to the bottom of the shaft is 100 meters. After $x = 20$ minutes, the equipment will be 28 meters from the bottom of the shaft.

25. $y = 45x + 180$ The slope, $m = 45$, indicates he farms 45 gold coins an hour. The y-intercept, (0, 180), indicates the number of coins he had when he started farming. After $x = 15$ hours, Marco will have farmed the 855 gold he needs to buy the armor.

27. $y = -12x + 5$

29. $y = 3x - 5$

31. $y = -1.5x + 33$

33. $y = 1.2x + 6$

35. The y-intercept is (0, 120) and indicates that Freddie initially had $120. The x-intercept is (8, 0) and indicates that he played 8 losing rounds and lost all of his money.

37. (a) The slope, $m = -12$, indicates that the water is draining from the trough at a rate of 12 gallons per minute. The y-intercept (0, 350) indicates there were initially 350 gallons of water in the trough.

   b) 182 gallons

   c) When $y = 100$, $x = 20.8\overline{3}$ After approximately 20.8 minutes there will be 100 gallons of water remaining.

   d) When $y = 0$, $x = 29.1\overline{6}$ It will take a little more than 29 minutes before the trough is empty.

1.4 USING LINEAR FUNCTIONS TO MODEL DATA

1. Interpolation is when you make a prediction within the domain and range of the given data.

3. Positive linear correlation means the data presents a nearly linear model with a positive slope (an increasing function). Negative linear correlation means the data presents a nearly linear model with a negative slope (a decreasing function).

5. Linearly related with a negative correlation.

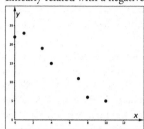

7. Linearly related with a positive correlation

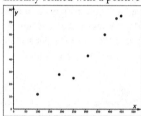

9. Equations will vary slightly depending on points chosen. One possible solution is $y = 112.5x + 11500$. In 2018, $x = 28$, we predict the population of Midgar will be 14,650. This is extrapolation.

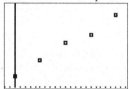

11. Graph 24d

13. Graph 24a

15. $y = 1.7x + 124$ In 2012, $x = 22$: the number of people in the U.S. labor force will be 161.4 million. In 2020, $x = 30$: the number of people in the U.S. labor force will be 175 million.

17. $y = 0.476x + 20.745$; When $y = 35$, $x = 29.947$; In about 30 years (2020) the percentage of persons 25 yrs or older who are college graduates will exceed 35%.

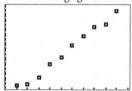

19. $y = -22.803x + 736.727$. In 2016, $x = 26$ and $y = 143.849$. The model predicts there will be about 144 people unemployed in Waldorf in 2016.

21. $y = 1.273x + 40.944$; The model predicts that in 2020, $x = 30$, about 79% of H > S. students will take Algebra 2.

23. $y = 0.066x - 7.149$; A sandwich with 620 calories will have about 33.8 grams of fat.

1.5 FUNCTION NOTATION AND MAKING PREDICTIONS

1. $-11, -1, 2a + 3$

3. $20, 15, 3a^2 + 20a + 32$

5. $k(2) = 3$, $t = 4$

7. a) 15
   b) $5, -5$

9. a) 5
   b) $x = 4$

11. a) 1,
    b) 6.5

13. a) -3
    b) $x = 0$, $x = -6$

15. a) 53
    b) $x = 2$

17. 8, 6, 4, 2

19. 21, 11, 3, -3

21. $-4, \frac{-3}{2}, \frac{-2}{3}, \frac{-1}{4}$

23. a) $f(40) = 13$, b) A population of 5,000 people produces 2 tons of garbage per week.

25. a) There are 30 ducks in the lake in 1995.
    b) There are 40 ducks in the lake in the year 2000.

27. a) $W(t) = \frac{1}{2}t + 7.5$
    b) Domain: [0, 1020] Range: [7.5, 150]

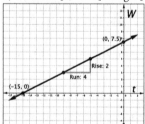

c) The W-intercept is Arlo's weight at birth. The t-intercept is when Arlo will weigh 0 pounds.

d) The slope of $\frac{1}{2}$ means that Arlo gains half a pound a month.

e) 5.8 months.

f) 10.6 pounds.

Chapter 2: Exponential Functions

2.1 PROPERTIES OF EXPONENTS

1. No, the two expressions are not the same. An exponent tells how many times you multiply the base. So 2^3 is the same as $2 \times 2 \times 2$, which is 8. 3^2 is the same as 3×3, which is 9.

3. Scientific notation is a method of writing very small and very large numbers.

5. 81

7. 243

9. $\frac{1}{16}$

11. $\frac{1}{11}$

13. 1

15. 4^9

17. 12^{40}

19. $\frac{1}{7^9}$

21. 3.14×10^{-5}

23. 16,000,000,000

25. a^4

27. $b^6 c^8$

29. $a b^2 d^3$

31. m^4

33. $\frac{q^5}{p^6}$

35. $\frac{y^{21}}{x^{14}}$

37. 25

39. $72\,a^2$

41. $\frac{c^3}{b^9}$

43. $\frac{y}{81\,z^6}$

45. 0.000022 m

47. 1.0995×10^{12}

49. 0.00000000003397 in.

51. 602214130000000000000000

53. $\frac{a^{14}}{1296}$

55. $\frac{n}{a^9 c}$

57. $\frac{1}{a^6 b^6 c^6}$

2.2 RATIONAL EXPONENTS

1. 2

3. 3

5. 16

7. 4

9. $\frac{1}{3}$

11. $\frac{1}{8}$

13. -4

15. -4

17. $3^{6/5}$

19. 320

21. 2

23. $\frac{1}{2}$

25. a^{35}

27. $\frac{1}{b^{1/2}}$

29. $5\,a^3$

31. $30ab^2$

33. $a^{1/2}$

35. $\frac{7}{b^{1/11}}$

37. $\frac{1}{9^{2/3}}$

39. $\frac{2b^{65}}{3c^{45}}$

2.3 EXPONENTIAL FUNCTIONS

1. For an exponential function, as the input values increase by 1, the output values are multiplied by the base.

3. The y-intercept occurs when $x = 0$. For $f(x) = ab^x$, $f(0) = ab^0 = a$, so the y-intercept is at $(0, a)$.

5. Exponential. The population decreases by a proportional rate.

7. Not exponential. The charge decreases by a constant amount each visit, so the statement represents a linear function.

9. The forest represented by the function $B(t) = 82(1.029)^t$.

11. After $t = 20$ years, forest A will have 43 more trees than forest B.

13. — Answers will vary, but here is a sample response: For a number of years, the population of forest A will increasingly exceed forest B, but because forest B actually grows at a faster rate, the population will eventually become larger than forest A and will remain that way as long as the population growth models hold. Some factors that might influence the long-term validity of the exponential growth model are drought, an epidemic that culls the population, and other environmental and biological factors.

15. Exponential growth. The growth factor, 1.06, is greater than 1.

17. Exponential decay. The decay factor, 0.97, is between 0 and 1.

19.

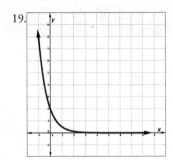

y-intercept: (0, 2)

21. y-intercept: (0, 5)

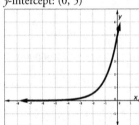

23. 20.736

25. 6.944

27. 28.739

29. 50.526

31.

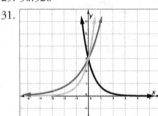

33. B

35. A

37. E

39. D

41. C

2.4 FINDING EQUATIONS OF EXPONENTIAL FUNCTIONS

1. $a = 8, b = 2, f(x) = 8(2)^x$
3. $m = -7, b = 11, f(x) = -7x + 11$
5. $a = 243, b = \frac{1}{3}, f(x) = 243\left(\frac{1}{3}\right)^x$
7. $b = \pm 7$
9. $b = 3$
11. $b = -0.5$
13. $b \approx \pm 2.667$
15. There are no real number solutions.
17. $b \approx 0.839$
19. $a = 6, b = 5, y = 6(5)^x$
21. $a = 450, b \approx 0.447, y = 450(0.447)^x$
23. $a = 4.5, b \approx 1.200, y = 4.5(1.2)^x$

25. $a = 11.77, b \approx 0.750, y = 11.77(0.75)^x$
27. Let $f(t)$ be the population of chicken turtles t years after the initial release.
    a) 0, 300) and (7, 550)
    b) $f(t) = 300(1.09)^t$
    c) $f(12) = 300(1.09)^{12} \approx 844$ Twelve years after the initial release, the chicken turtle population will be about 844.
29. Let $f(t)$ be the value of the Prius t years after 2007.
    a) 0, 19500) and (9, 5600)
    b) $f(t) = 19500(0.871)^t$
    c) $f(13) = 19500(0.871)^{13} \approx 3237.95$ In 2020, the value of the car will only be about $3238.

2.5 USING EXPONENTIAL FUNCTIONS TO MODEL DATA

1. exponential growth, 19% increase per unit of time
3. exponential decay, 2% decrease per unit of time
5. exponential growth, 2.85% increase per unit of time
7. exponential decay, 50% decrease every 12 units of time (or approximately 5.61% decrease per unit of time)
9. $f(t) = 42,000(1.0254)^t$
11. $f(t) = 250\left(\frac{1}{2}\right)^{t/28}$ or $f(t) = 250(.9755)^t$
13. $f(t) = 1500(2)^{t/12}$ or $f(t) = 1500(1.0595)^t$
15. $f(t) = 2340(0.905)^t$

17. a) $f(t) = 2700(1.0325)^t$
    b) In 5 years, about $3168.21, and in 10 years, about $3717.61.
19. a) $f(t) = 1000(2)^{t/7}$ or $f(t) = 1000(1.1041)^t$
    b) In 25 years, about $11,887.95.
21. a) $P = f(t) = 9740(0.928)^t$
    b) About 712 people in 2015.
23. a) $V = f(t) = 32000(0.76)^t$
    b) In 4 years, about $10,676.

25. a) $f(t) = 40\left(\frac{1}{2}\right)^{t/1620}$ or $f(t) = 40(.9997)^t$

b) After 2000 years, 17 mg.

27. $200\left(\frac{1}{2}\right)^{12020} = 3.125$mg or $f(t) = 200(.9659)^t$

29. a) $y = 84.638(1.012)^x$

b)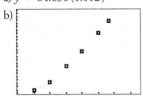

c) 1.2% increase per year.

d) In 2025, about 375.1 million people, but the answers may vary slightly due to rounding.

31. a) $y = 0.690(1.084)^x$

b)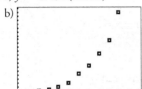

c) Since $a \approx 0.690$, we estimate that the revenue collect by the IRS in the initial year 1900 (in this problem) was 0.690 billion dollars or about $690,000,000. Since $b \approx 1.084$, the percent change is 8.4% increase per year.

d) In 2018, the IRS will collect about 9382.2 billion dollars or about $9,382,000,000,000.

33. a) $y = 402.388(2.283)^x$

b)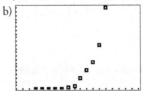

c) In 2000, there were approximately 504,055,912 computers connected to the Internet, but answers may vary due to rounding.

Chapter 3: Logarithmic Functions

3.1 LOGARITHMIC FUNCTIONS

1. 3	15. –2	29. ¼	43. 30
3. 5	17. –5	31. 5	45. 5
5. 2	19. 0	33. 3.831	47. 6
7. 4	21. 1	35. 1.462	49. 4
9. 7	23. ½	37. –0.604	
11. 0	25. ⅓	39. 4	
13. –3	27. ½	41. –2	

3.2 PROPERTIES OF LOGARITHMS

1. $4^3 = 64$	21. 1	41. 7.2380
3. $5^4 = 625$	23. $\frac{1}{1000} = 0.001$	43. 2.7030
5. $2^{-3} = \frac{1}{8}$	25. 4	45. 0.9783
7. $10^0 = 1$	27. 25	47. 9.2788
9. $w^r = m$	29. 2	49. 6.8480
11. $\log_2(256) = 8$	31. 7	51. $x = -25, x = 1.1946$
13. $\log_{49}(7) = \frac{1}{2}$	33. 81	53. 1.5150, 4.9530
15. $\log(10,000) = 4$	35. 2.1129	55. no real-numbered solutions
17. $\log_4(97) = x$	37. 4.2009	
19. 144	39. 3.2161	

57. a) $f(t) = 2200(1.047)^t$
    b) $3817.53
    c) 17.9 years
59. in 23.5 years
61. a) $f(t) = 15,000(0.96)^t$
    b) 9972 whales
    c) in 15.4 years
63. in 4.7 years

65. a) 180 mg
    b) After 12 hrs there are 45 mg of caffeine remaining.
    c) in 25 hrs.
67. a) The population is increasing by 1.3% per year.
    b) In 2008, the population was 8.649 thousand people.

c) in 2019
69. a) about 680.3 million airline travelers
    b) in 2032
71. in 2021
73. 9 years

3.3 Using Logarithms to Make Predictions with Exponential Models

1. 2.3823
3. 8.4338
5. 0
7. 12
9. –4
11. $\frac{1}{2}$
13. 9
15. $x = 54.5982$

17. $x = 11.3891$
19. $x = 24.7355$
21. $x = 4.8866$
23. $x = 5.7323$
25. $x = 7.8396$
27. $x = 5.1358$
29. a) There were initially 500 bacteria.
    b) about 725 bacteria (c) 14.1

hours
31. a) The initial investment was $15,000.
    b) $23,290.61 (c) 22.3 years
33. 5.2 minutes
35. 12.9 years
37. 47.8 years

Chapter 4: Quadratic Functions

4.1 Expanding and Factoring Polynomials

1. $48x^7y^8$
3. $-14a^5b^7$
5. $4x^3 + 36x^2 - 20x$
7. $-21x^4 + 33x^3 - 6x^2$
9. $-18a^3b^2 + 15a^2b^2 - 12ab^2$
11. $2x^4y + 18x^3y^2 - 2x^2y^3$
13. $x^2 - 2x - 8$
15. $x^2 + 19x + 88$
17. $x^2 - 15x + 54$
19. $x^3 + 8x^2 - 4x - 32$
21. $8a^4 - 4a^3 + 24a^2 - 12a$
23. $20x^5 + 4x^4 - 15x^3 - 3x^2$
25. $12a^2 - 17ab - 5b^2$
27. $x^2 + 18x + 81$
29. $a^2 - 14a + 49$
31. $4m^2 - 12m + 9$
33. $x^2 - 4$
35. $b^2 - 36$
37. $16x^2 - 1$

39. $4a^2 - 49b^2$
41. $2x(7 + 2y - 9y^2)$
43. $15xy(2x^2 - 3xy + 9y^2)$
45. $18j^2k^2(2j^2 - jk + 3k^2)$
47. $16r^3(5r^4 - t)$
49. $(a + 11)(a - 2)$
51. $(x + 5)(x + 7)$
53. $(x - 8)(x - 1)$
55. $(t - 12)(t + 3)$
57. $(7x - 1)(x + 7)$
59. $(2b - 3)(b + 8)$
61. $(5t - 4)(t - 3)$
63. $(x + 1)(x - 1)$
65. $(2m + 3)(2m - 3)$
67. $(5x + 14)(5x - 14)$
69. $(12a + 7b)(12a - 7b)$
71. $x = 0, x = -8$
73. $a = 0, a = 7$
75. $x = 11, x = -11$

77. $m = 10$
79. $x = \frac{5}{2}, x = -\frac{5}{2}$
81. $x = 6, x = 3$
83. $x = -7, x = 5$
85. $x = -\frac{2}{3}, x = -\frac{8}{3}$
87. $x = \frac{1}{2}, x = \frac{-5}{5}$
89. $x = 0, x = \frac{5}{4}$
91. $x = 3, x = -2$
93. $x = 6, x = -2$
95. $x = -5, x = 5$
97. $x = 0, x = -1, x = -4$
99. $x = 0, x = -4, x = -11$

4.2 QUADRATIC FUNCTIONS IN STANDARD FORM

1. a) opens up
   b) table

x	–3	–2	–1	0	1	2	3	4
$f(x)$	4	–5	–8	–5	4	19	40	67

   c) Not all points from the table are on the graph.

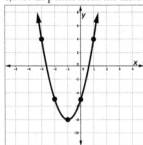

   d) (–1, –8)
   e) (0, –5)

3. a) opens down
   b) table

x	–3	–2	–1	0	1	2	3	4
$f(x)$	–18	–9	–2	3	6	7	6	3

   c) Not all points from the table are on the graph.

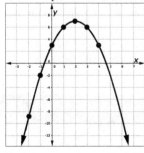

   d) (2, 7)
   e) (0, 3)

5. a) opens up
   b) (0, –4)
   c) (0, –4)

7. a) opens up
   b) (0, 13)
   c) (–4, 3)

9. a) opens down
   b) (0, 5)
   c) (1, 6)

11. a) opens up
    b) $(0, \frac{5}{4})$
    c) $(\frac{1}{2}, 1)$

13. vertex: (–1.5, 7); Domain: all real numbers; Range: $y \leq 7$

15. vertex: (–1, 5.6); Domain: all real numbers;
    Range: $y \leq 5.6$

17. vertex: (3, –9); Domain: all real numbers; Range: $y \geq -9$

19. vertex: (–2, 4); Domain: all real numbers; Range: $y \geq 4$

21. a) $A = -2W^2 + 130W$
    b) $W = 32.5$ ft.
    c) The maximum area is 2112.5 sq. ft.

23. The manufacturer should produce 20 garden hoses per day for a minimum production cost of $700.

25. 16 feet

27. $x = 30$; Producing 30 units per day will yield a maximum profit of $1000.

29. $P = -20n^2 + 200n + 97,500$; $n = 5$; Producing 5 more trees will yield a maximum crop of 98,000 apples.

4.3 THE SQUARE ROOT PROPERTY

1. 14

3. $2\sqrt{7}$

5. $6\sqrt{5}$

7. $\frac{4}{9}$

9. $\frac{\sqrt{17}}{12}$

11. $\frac{6\sqrt{5}}{5}$

13. $\frac{\sqrt{7}}{7}$

15. $\frac{\sqrt{21}}{7}$

17. ±6

19. $\pm\sqrt{14}$

21. $\pm 2\sqrt{10}$

23. $\pm\frac{\sqrt{2}}{3}$

25. ±2

27. $\frac{1}{3}$, 5

29. $-6 \pm \sqrt{7}$

31. $-9 \pm 2\sqrt{3}$

33. –8, 10

35. $3i$

37. $5i\sqrt{2}$

39. $\frac{2}{5}i$

41. $8i\sqrt{3}$

43. ±2i

45. $\pm3i\sqrt{5}$

47. $\pm i\sqrt{2}$

49. $9 \pm 10i$

51. $-1 \pm 2i\sqrt{2}$

4.4 THE QUADRATIC FORMULA

1. $4, -\frac{1}{2}$

3. $\frac{3 \pm 2\sqrt{6}}{5}$

5. $\frac{1 \pm i\sqrt{11}}{6}$

7. $-\frac{4}{3}$

9. $(-1, 0), (-1.5, 0)$, and $(0, 3)$

11. $(-3.225, 0), (-0.775, 0)$, and $(0, -5)$

13. $(4.449, 0), (-0.449, 0)$, and $(0, -2)$

15. $(0.425, 0), (-1.175, 0)$

17. $(0.563, 0), (9.237, 0)$

19. $x \approx 1.653, x \approx -6.653$

21. $t \approx 4 \rightarrow$ April 1999, $t \approx 24 \rightarrow$ Dec. 2000

23. 2.727 secs.

25. discriminant = 73; two real-number solutions; two x-intercepts

27. discriminant = –92; two imaginary-number solutions; no x-intercepts

29. discriminant = 0; one real-number solution; one x-intercept

31. discriminant = 1; two real-number solutions; two x-intercepts

33. $5 \pm 6i$

35. $2, -9$

37. $-0.208, 3.208$

39. $1, -\frac{1}{4}$

41. $-3 \pm i\sqrt{3}$

43. $1.589, -1.743$

45. $\pm i\sqrt{7}$

47. $-6, 5$

49. $-2, 5$

4.5 MODELING WITH QUADRATIC FUNCTIONS

1. $R = -0.0125x^2 + 45x$

   1800 thousand phones (or 1,800,000 phones)

   40500 thousand dollars (or $40,500,000)

3. 2.449 secs; 37.388 meters

5. 2100 rpm; 850 horsepower

7. a) At t = 0 sec, the softball is thrown from a height of 6.5 ft, and at 2 secs, the height is 32.5 ft

   b) It takes 1.406 sec to reach a maximum height of 38.14 ft

   c) 2.95 sec

   d) 0.5 and 2.313 seconds

9. a) $y = 48.81x^2 + 25x - 181$

   b) about 11,179 kg

11. a) $y = .05x^2 + 4.55x - 32.15$

    b) $47.48 \times 100 = 4748$ units sold.

    c) $x \approx 7.693$ and $x \approx 87.078$. These are the break–even values. The company will make a profit if it sells between 769 and 8708 units.

13. a) linear: $y = 1.753x + 14.7$, quadratic: $y = 0.097x^2 + 0.788x - 16.1$

    b) The quadratic model appears to be the better fit because the data points are closer to the curve when we graph the equation and the scatter plot in the same window.

    c) linear: $60.278 billion; quadratic: $102.16 billion; The linear model may give the more reasonable prediction.

Chapter 5: Further Topics in Algebra

5.1 VARIATION

1. $y = 20x$

3. $p = 3q^3$

5. $V = 6t^4$

7. $y = \frac{8}{x}$

9. $h = \frac{1.28}{d^2}$

11. $r = \frac{15}{\sqrt{A}}$

13. $y = 34$

15. $t = 3$

17. $y = 8$

19. $t = 2$

21. Q increases

23. T increases

25. less resistance

27. 3 seconds

29. 48 inches

31. 49.75 pounds

33. $C = 33.33$ amperes

5.2 ARITHMETIC SEQUENCES

1. A sequence where each successive term of the sequence increases (or decreases) by a constant value.

3. We find whether the difference between all consecutive terms is the same. This is the same as saying that the sequence has a common difference.

5. 17, 29, 41, 53, 65; arithmetic with $d = 12$

7. $\frac{1}{2}, \frac{1}{3}, \frac{1}{4}, \frac{1}{5}, \frac{1}{6}$; not arithmetic

9. -2, 13, 28, 43, 58; arithmetic with $d = 15$

11. 99, 96, 91, 84, 75; not arithmetic

13. arithmetic; $d = \frac{1}{2}$

15. The sequence is not arithmetic because $16 - 4 \neq 64 - 16$.

17. arithmetic; $d = -21$

19. arithmetic; $d = -1.9$

21. $a_n = 5n + 3$; $a_{25} = 128$

23. $a_n = -7n + 34$; $a_{25} = -141$

25. $a_n = -25n + 13$; $a_{25} = -612$

27. $a_n = 1.75n + 21.75$; $a_{25} = 65.5$

29. $n = 20$

31. $n = 238$

33. $n = 82$

35. $n = 35$

37. No; $n \approx 25.3$ which is not a positive whole number.

39. a) $a_n = 3n + 18$

b) $a_{12} = 54$, There are 54 seats in the 12th row.

c) 40 rows

41. a) $a_n = 950n + 31,550$

b) $a_5 = 36,300$, She will make $36,300 in the 5th year of employment.

c) in the 20th year

43. a) $a_n = -8n + 217$

b) $a_{14} = 105$, After 2 weeks (14 days) there are 105 food units left.

c) 27 days

5.3 GEOMETRIC SEQUENCES

1. A sequence in which the ratio between any two consecutive terms is constant.

3. Both geometric sequences and exponential functions have a constant ratio. However, their domains are not the same. Exponential functions are defined for all real numbers, and geometric sequences are defined only for positive integers.

5. geometric; $r = 3$

7. arithmetic; $d = -20$

9. neither

11. geometric; $r = \frac{1}{2}$.

13. arithmetic; $d = 9$

15. neither

17. $a_n = 7(4)^{n-1}$

19. $a_n = \left(\frac{4}{5}\right)^{n-1}$

21. $a_n = 768\left(\frac{1}{4}\right)^{n-1}$

23. $a_n = 10(7)^{n-1}$

25. 196,608

27. 4.768×10^{-6}

29. 9.038×10^{11}

31. $n = 12$

33. $n = 8$

35. $n = 16$

37. $n = 15$

39. a) $a_n = 20\left(\frac{7}{8}\right)^{n-1}$

b) 6.872 ft.

40. a) $a_n = 8\left(\frac{4}{5}\right)^{n-1}$

b) 1.074 m

41. $5,368,709

43. a) $a_n = 500\left(\frac{2}{3}\right)^{n-1}$

b) $a_4 = 148.148$; About 148 cubic cm of air remain after 4 cycles.

c) 13 cycles

45. a) Aaron: $a_n = 40000(1.0375)^{n-1}$
Beatriz: $a_n = 1000n + 44000$

b) Aaron: $a_4 = $44,671$,
$a_{12} = $59,969$
Beatriz: $a_4 = $48,000$,
$a_{12} = $56,000$

c) Aaron: $a_{30} = $116,336$
Beatriz: $a_{30} = $74,000$

d) year 9

CPSIA information can be obtained
at www.ICGtesting.com
Printed in the USA
FSOW04n0535261116
27675FS

9 781943 536139